ATLANTIC STUDIES ON SOCIETY IN CHANGE

NO. 129

Editor-in-Chief, Béla K. Király

Associate Editor-in-Chief, Kenneth Murphy

Editor, László Veszprémy

War and Society in East Central Europe
Volume XLI

THE HISTORY OF THE HUNGARIAN MILITARY HIGHER EDUCATION 1947-1956

Miklós M. Szabó

Translated by Nora Arato and Brian McLean
Copyedited by Matthew Suff
Typeset by Andrea T. Kulcsár

Social Science Monographs, Boulder, Colorado
Atlantic Research and Publications, Inc.
Highland Lakes, New Jersey

Distributed by Columbia University Press, New York
2006

EAST EUROPEAN MONOGRAPHS, NO. DCCIII

The publication of this volume was made possible by grants from
Prime Minister's Memorial Committee

Publication of the Hungarian Academy of Sciences

Copyright © 2006
by Atlantic Research and Publications, Inc.

ISBN 0-88033-601-3
Library of Congress Catalog Card Number 2006933155

Printed in Hungary

Table of Contents

List of Pictures

Preface

The volume deals with the higher education of Hungarian officers in an era of substantial and tragic changes in the political system of the country. Hungary was "liberated" from the Nazis by the Soviet armed forces. Alas, the liberation turned out to be the imposition of a Bolshevik system, dominated by the Soviet Union, upon the country, just as it was upon all the East Central European states except for Austria.

A point of the Yalta agreement guaranteed the sovereignty and independence of the liberated countries (former allies and enemies of the Grand Alliance alike)—namely the holding of general elections by secret ballot. This was honored by the Soviet Union in Hungary. The provisional government (December 22, 1944–November 15, 1945) was replaced by a democratically elected parliament. The Soviet leaders expected the Hungarians to express their gratitude for the "liberation" by electing a Communist-dominated legislative body. The Soviet occupation authorities did not interfere with the elections, and thus a truly clean general election was held, by secret ballot. All the pre-World War II democratic political parties, plus the Communist Party (banned prior to 1944), participated.

The elections were held on November 4, 1945. The Independent Small-holders', Agrarian and Citizens' Party won an absolute majority, with 57 percent of the votes. The Small-holders, together with the other democratic parties, had an overwhelming majority in the Parliament. The Communist Party won 17 percent of the votes, and was thus a negligible force in the legislative body. However, until the post-World War II Paris Peace Treaty came into force on September 15, 1947, supreme political power was exercised in Hungary by the Allied Control Commission. Its president,

Marshal of the Soviet Union Kliment Yefremovich Voroshilov, did not permit the formation of a government without key ministries being given to the Communists. In every respect, particularly as regards the armed forces, Voroshilov's decision was final: all appointments, dismissals and promotions, and deployment of the troops, to mention only a few such matters, needed his approval.

Voroshilov used this authority to the full, ensuring that before September 1947 in Hungary political power was in the hands of the Communist Party (Workers' Party). The country remained occupied by Soviet troops, who were there in principle to secure the connection between their homeland the Soviet Union and the troops occupying Austria. They had no right to interfere with Hungary's domestic affairs, but they did, supporting the Bolsheviks' consolidation of power.

A hope prevailed that after the Austrian state treaty was signed the Soviet troops would leave both Austria and Hungary. However, a day before the Austrian state treaty was signed, on May 15, 1955, the Warsaw Pact was forced upon Hungary and all her neighbors (except for Austria and Yugoslavia). Thus the Soviet occupation of Hungary continued throughout the entire period that this volume covers. This must be emphasized, in order that the immense data— collected by intensive research on the part of the author—may be comprehensible.

With these general considerations in mind, it seems proper to point out some of the most important elements of the subject. A substantial part of this volume deals with the era between 1948 and 1953, during which the Hungarian army became much larger than it had ever been before in peacetime. This was due to the Stalin–Tito split, which meant that Stalin was ready and willing to destroy Tito by all means, including war. In that conflict the captive nations of the Soviet colonial empire were supposed to participate with armed men and industrial production, as well as by "cleansing" their societies of all national characteristics and traditions and establishing a Bolshevik society, a carbon copy of the Soviet system. The basic political-ideological aim, absolute loyalty to Stalin, Bolshevism and the Soviet Union, demanded the elimination from the officer corps of all persons whose dedication to their nation—in other words their patriotism—made them suspicious for the Bolsheviks. This "cleansing" process was gradual. As long as their professional knowledge was indispensable, they were tolerated; hundreds of them were even recalled for active service from unemployment. As soon as their service was thought to be no longer indispensable, they were dismissed; many of them were jailed, others physically eliminated. The most sensational of these actions occurred in 1949, when Inspector-General György Pálffy was executed, and in 1950, when Lieutenant Generals László Sólyom and Gusztáv Illy and four of their fellow generals met the same end. This wave of terror was not publicized, however: those who were in danger knew it very well in any case.

Since in principle as well as in practice loyalty to Bolshevism overruled professional knowledge, the dismissed or eliminated well-trained officers were replaced by individuals the professional knowledge of the utmost majority of whom was zero, and whose general educational level was that of elementary school. Their social origins were predominantly industrial workers and poor peasants. It was unprecedented in the Western world that persons with such a minimal level of preparedness represented the utmost majority of those educated and trained at college and university level.

To reduce these shortcomings, and also to secure direct Soviet control of the captive nations' armies, from the fall of 1948 onwards Soviet military "advisers" were assigned to all command headquarters from the Ministry of Defense down to the battalion command level. At the Military Academy every faculty or department had at least one tactical and one technical adviser. They were well-educated General Staff officers with substantial combat experience as World War II veterans.

A technical problem made the communication between the advisers and the Hungarian personnel difficult: the Hungarians' poor knowledge of the Russian language, despite the intensive compulsory Russian-language studies. For each Soviet adviser an interpreter was assigned. Here again the question of loyalty overruled the knowledge of either language. Therefore most of the interpreters were selected from among the Sub-Carpatho-Russian province of Ukraine. In principle they were well versed in both languages, but in practice they were not. In addition, their military education was zero or was at the level of a common soldier, when their job was translation of exchanges on strategy and higher leadership in general.

In the Korean War, starting in 1950, the general expectation that the Western nations (the USA in particular) were "paper tigers"—which meant that they would not interfere if the Communist Great Powers (China and the Soviet Union) expanded their sphere of influence by military action—was proven false. Stalin for his part did not wish to cross swords with the West, and he therefore gave up the idea of removing Tito by means of arms. But that change of fundamental policy was not made public until Stalin died, on March 5, 1953.

From 1950 onwards the possibility of war evaporated, yet the captive nations had to maintain their immense armed forces for nothing. Only after Stalin's death did the new Soviet leadership openly give up hostile policies against Tito and permit the captive nations gradually to reduce their huge armed forces. Year after year, substantial reduction of the armies' strength meant for the officer corps that their employment became insecure: many of them lost their jobs, while the remainder were often transferred to other units and had to change their place of residence. This reduced the spirit of the officer corps. By 1956, at the outbreak of the Hungarian

Revolution, the Hungarian armed forces were commanded by a demoralized officer corps.

All these factors made the operation of the higher officers' training system confused and poorly organized; the quality of education and that of the graduating officers were greatly below requirements.

Some elements of this ineffective system were the following: constant interference with the higher officers' educational institutions' everyday life, frequent changes of organization and of the annual curriculum, and an immense amount of compulsory periodic reports on the status of education, morale, the achievements of students and the like. In short, instead of the assignation of long-term tasks, the institutions received permanent interference of a bureaucratic nature rather than high-level orientations.

The following authorities interfered frequently with the affairs of schools: the Ministry of Defense, the General Staff, the Political Chief Section of the Ministry of Defense, and the Branch Commands; the infantry, armored troops, artillery, air force, engineering troops and signals troops each had a Branch Command under the leadership of a general directly subordinated to the minister of defense.

If the present study seemingly deals too much with these problems, this is not a mistake but the merit of the author, because it reveals the system as it was organized and how it operated.

Budapest, the fiftieth anniversary of   Béla K. Király
the Revolution and War for Independence   Editor-in-Chief

The main building of the Royal Hungarian Military Academy in 1940
Inventory number: 62264/alb

Introduction

Based on available documents and resources I have tried in this book to give a complete overview of the activities and history of the *Honvéd* Military Academy, the *Honvéd* Academy, and the political officer-training academies. In addition, I have included the first 1.5 years' history of the Miklós Zrínyi Military Academy—which has been functioning under this name since March 15, 1955, to honor the approaching fiftieth anniversary of the 1956 Hungarian Revolution and War of Independence and to honor this military educational institution's important role in the military events in Budapest during the Revolution.

The intense struggle to establish and develop high-level officer training in Hungary after World War II took its toll among the victims, who sacrificed their lives or livelihood in participating in this historical mission. It also took much material, moral and health sacrifice from those who became dedicated to this cause but who, due to their low level of educational preparedness, were consumed during this superhuman ordeal. Yet, after this huge sacrifice, an increasing number of people at an increasing level were able to meet the demands of growing expectations and laid the foundations of a state-of-the-art military higher education within hardly a decade.

Thanks to all who helped improve this book with their advice and recommendations.

Chapter I
The History of the *Honvéd* Military Academy
(December 1, 1947–January 17, 1949)

Historical Antecedents

In the Austro-Hungarian Monarchy, Hungarian General Staff officers acquired the necessary qualifications at the *Kriegsschule* in Wiener Neustadt and at the information courses of Belgrade and Laibach during World War I. Following the disintegration of the Monarchy—and the effects of the Peace Treaty of Trianon—the Hungarian General Staff officer training began secretly at the Hungária körút [boulevard] base of Budapest on September 27, 1920. After resorting to several codenames, the base could finally and openly adopt the name *Hungarian Royal* Honvéd *Military Academy* in 1939.

Due to the military developments, in July 1944 the educational institution resettled in Balatonfüred, where it was disbanded on October 17. Yet the second-year students were ordered to go to Szombathely at the end of December; there their education continued until early February 1945, at which point they embarked on a three-month-long study tour in Germany. From Germany they were unable to return home, and they stayed in the mountain infantry barracks of Lenggries until the end of the war.[1] Thus the first period of the Hungarian General Staff officer training ended.

The Reorganization of the Military Higher Education

After the immense war losses, about 60 percent of the officer corps was dismissed between May 1945 and October 1946, and the number of officers dismissed exceeded 6,700 by the end of 1947.[2] This

1

extraordinary loss of officers—including a high number of General Staff officers—was impossible to compensate by replacements, since, with the exception of certain short courses, between 1945 and 1947 no military institution of any level functioned in Hungary.

Yet the gradual unfolding of the Cold War and the fact that the Left was gradually gaining ground put the revival of different military institutes on the agenda. Accordingly, in the spring of 1947 the establishment of a new type of General Staff officer corps was acknowledged as a vital need. The appeal that called eligible officers to apply for General Staff officer training served this purpose.[3] The requirements were specified in the general order issued on April 10, 1947.

In the general order the following essential decision was spelled out for the first time: "In the following training year in all probability the *Military Academy* will be established (in Budapest) and the General Staff officer training will be launched." Professional officers were allowed to apply until May 10 through the official channels if they were under 35 years of age, had at least three years of military experience (acquired by October 1, 1947), and held a rank not higher than that of a major. "The autobiography written according to the attached sample should be attached to the application, along with the certification of graduation from a military or public school, as well as the certification form acquired from the relevant screening committee by higher commands."

The fact that in the given historical period only the cream of the officers of the former Hungarian Royal *Honvédség* could meet the extraordinarily high admission requirements gave special significance to the certification process of the screening committees.[4] General cultural and political knowledge, history (specifically modern times and the history of Hungary), battalion and regiment tactics, general world geography and detailed geography of neighboring countries, knowledge of infantry and artillery firearms, and map-reading were subjects of the entrance examination. The general order said the following: "I expect that all the professional officers who meet the above requirements and feel a calling for higher training will apply in high numbers. The applicants should utilize the remaining time to expand the necessary knowledge and present themselves at the entrance examination with conscientious preparation and thorough knowledge."

This system of requirements—which, for the insiders of the topic, is obviously the modernized and changed version of the experiences of the Hungarian Royal *Honvéd* Military Academy, adjusted to the changed political situation—was developed by Major General László Sólyom, head of the Chief Section for Operations of the Ministry of Defense, in early 1947. Its essence was that the most qualified 15–20 young officers of adequate troop experience would be prepared to make excellent General Staff officers and would later cope with higher leadership assignments.[5]

350 officers applied to the program after the general order was issued, and 70 of them were called in for the time period of September 5–15 to take written and oral examinations, which took place both in classrooms and in the field. As a result, 18 of them were admitted, of whom three became better known later: Major Pál Demtsa (major general), Lieutenant Lajos Móricz (colonel, doctor of military sciences), and Captain Miklós Szűcs (colonel, operational sub-department head).[6]

After the publication of the quoted general order of March—which had spoken of the establishment of the Defense Academy—several interesting events happened in Hungarian military policy. Although the Military Sub-committee of the Hungarian Communist Party agreed with the plan, the Allied Control Commission working in Hungary did not agree to the Academy's adopting this name. However, it did agree to establish the new educational institution as an Officer Further Training Course. This situation is reflected by the 1947 (July 11) provision[7] of the Ministry of Defense, which said the following: "In the organizational order of the 1947–1948 academic year... I will, in detail, dispense for the... staff... of the Officer Further Training Course."

There is some uncertainty both in the professional literature and in the documents themselves regarding the restoration of the name of the Defense Academy. According to some experts,[8] since the Allied Control Commission ceased to exist with the Paris Peace Treaty (September 15, 1947) coming into force, reverting to the name Defense Academy became possible. Yet the 1947–1948 organizational order[9] issued on July 23, 1947, seemed to contradict this by declaring its intention "to establish the *Honvéd* Military Academy (former Officer Further Training Course)." But another disposition of October 21, 1947,[10] still spoke about the Officer Further Training Course.

The Appendix 45 of the organizational order contained the "Organizational and Staff Chart of the *Honvéd* Officer Further Training Course." Its original version spoke about 35 professional officers with 30 students among them, 13 non-commissioned officers, and 33 serving in the ranks, that is, 81 soldiers. (Later notes and corrections showed a decrease of almost 50 percent in the case of the enlisted.) The commander was a general.

As a temporary solution, the head of the Financial Sub-department of the Ministry of Defense ordered that "the Military Academy is to be established on September 15 of this year."[11] Obviously, it was not their fault—and the circumstances were to be blamed—that the beginning of the year had to be postponed several times.

Although one of the publications indicated that on October 1, 1947, "the *Honvéd* Military Academy as an institution of higher-level officer training is to be established,"[12] Major General Sólyom's measure of October 21[13] contradicted it by saying the following: "The date of opening of the Officer Further Training Course is uncertain for the time being, due to unfinished construction work. As far as can be foreseen, the intake will happen during the first week of November." He also permitted a vacation for the students, to be spent at their places of domicile, from November 1. The paragraph of the government decree included the stipulation that the students "should possibly bring along" compasses, tele-scopes, caliper compasses, drawing boards, ink, pens, and colored and soft pencils, and regulations were indications of poor financial preparations.

Yet, at the same time, efforts were made to establish a high-level teaching faculty. For example, Major General Sólyom, head of the Chief Section for Operations of the Ministry of Defense and commander of the Academy, taught military strategy and military history (among other things), General Staff Colonel László Vértes taught military tactics, General Staff Colonel Sándor Lőrincz taught operational logistical service, and Major General László Réczey dealt with army organizations. General cultural subjects were taught by civilian university professors, for example, Tamás Nagy (political economics), László Bóka (literature), Erik Molnár (Hungarian ancient history), Aladár Mód (Hungarian history) and János Beér (state and legal studies).[14]

On behalf of the Ministry of Defense, Major General Gusztáv Illy issued an order on October 21, 1947, in which a provision was made to transfer the admitted students to the staff of the Headquarters of the Officer Further Training Course. Altogether one major, seven captains, nine first lieutenants and one second lieutenant expected to begin their studies.[15]

The general order issued on November 10, 1947, by the min-ister of defense finally ended the several-month-long name-feast by saying "I change the name of the 'Officer Further Training Course'…, which was established by my order, to *Honvéd* Military Academy."[16]

The Functioning of the *Honvéd* Military Academy

After several deadline modifications, finally December 1, 1947, the day of opening the *Honvéd* Military Academy, arrived. The signif-icance of the event was further emphasized by Prime Minister Lajos Dinnyés's and Minister of Defense Péter Veres's participa-tion at the long-awaited celebration.[17]

In his opening speech[18] Major General Sólyom stated that "the atmosphere of the school makes its mark on the spirit of the future *Honvédség* [Hungarian Army], and the horizons and professional knowledge that students acquire here in two years will define the military standards of the armed forces for a generation to come... and we lay *the foundation stone of a new General Staff* by opening the Military Academy..."

He also pointed out that the old General Staff could not have led the Hungarian military to the path of democracy and development. Hardly half a percent of the old General Staff officers still served, and that number did not endanger new ambitions. This was owed to the following: "The elimination of the past was not as radical in any areas of Hungarian public life as in the Hungarian military and, especially, in the General Staff." He was glad to mention the fact that "all of those whose opinions count *uniformly recognized* the necessity of the Hungarian military... Therefore, the support and development of the Hungarian military have become evident with the strengthening of the people's democracy. Furthermore, opening the Military Academy and beginning the General Staff officer training is an initiative of the utmost importance, with regard both to developing the Hungarian military and to further strengthening the Hungarian democracy."

Sólyom also outlined the role of the General Staff:

The armed forces are in need of a confident General Staff, that is, *leadership* with excellent professional expertise, which during peacetime ensures the spirit and the high-level expertise of the armed force and which, in the event that anybody endangers the freedom of the people, heads the armed force and commands it triumphantly... The new General Staff will not only be different from or better than its predecessors with regard to its place occupied in the Hungarian society and in the community of democratic peoples, but will surpass them in professional expertise as well.

Lieutenant General László Sólyom,
Chief of General Staff in 1949/50
Inventory number: 14893

Major General Sólyom's activities as "teacher of military history" also proved the efforts made to offer a comprehensive general educational background. During his classes "he often alluded to the military scientific works of Julius Caesar, Machiavelli, Miklós Zrínyi, Clausewitz, Delbrück and Engels,"[19] and was inclined to quote examples from World War II. This, however, seems to be contradicted by the statement that he "did not want to educate individuals who are 'strategist's geniuses' and 'individuals who are skilled at doing everything' but good division and army corps officers."[20] He also said the following: "I believe that the latter professionals also needed and still need comprehensive education, for two reasons: 1) Nobody from among the talented ones wishes to be stuck at the divisional or army corps staff level, and 2) It is difficult to evaluate General Staff officer students of mostly under 30 years of age and decide who will make a 'great military leader'."[21]

This work, which demanded high performance, could only be achieved by means of studying according to a stringent daily schedule and by applying discipline. Weekdays started with riding—except for the "horse rest" on Thursdays—and physical education on Thursdays. These activities were followed by classes between 9:00 a.m. and 1:00 p.m. One day a week was spent in the field. In the afternoons there were usually two-hour-long "horizon-expand-ing" lectures, debates, or driving. During the rest of the day the students had to prepare for the following day and to study another language, which was not compulsory, since one prerequisite of admission had been proficiency in one of the world languages.

László Sólyom, who was promoted to the rank of lieutenant general on April 4, 1948, evaluated and ranked the students every three months based on their achievements. The first four places of this scale of values were solidly and fully maintained by Miklós Szűcs, Pál Demtsa, Lajos Móricz and László Kalicza.[22]

Following the summer vacation in July the students were detailed to a month-long residential (Communist) Party school in order for them to understand the significance of the two (Communist and Social Democratic) parties' unification.[23] Following that, September was spent on a four-week-long tactical trip when regimental- and divisional-level issues were overviewed in the field by individual students giving military historical talks on different sites—talks that they prepared for by themselves—and the students participated in factory visits.[24]

It indicated a significant change in the Soviet Union's policy that following Péter Veres's resignation on September 9, 1948, Mihály Farkas, the deputy secretary-general of the Hungarian Workers' Party, was appointed as minister of defense.[25] The effect of this change was felt shortly after the opening of the second academic year on October 1. From then onwards the students of two years studied in the school's final location—in the buildings of the former Hungarian Royal *Honvéd* Military Academy in Hungária körút. The general order,[26] which was completely identical to that of the previous year, was issued already on March 19, 1948, and it contained the current yearly application to the Military Academy. There must have been some problem with the applications to the Academy—with possibly fewer "new cadres" among them—because according to the general order dated on May 12, "I will extend the deadline of applying to the Military Academy from May 1, 1948, to June 30, 1948."[27]

22 of the 66 applicants to the new class were successful, although only 20 of them started their studies.[28] The experience from the previous year was repeated, although in not such an overwhelming way: half of those admitted had graduated from the Ludovika or the Bolyai Military Academies, 2 of them had university diplomas, and 15 of them had passed exams in one or several languages.[29]

The effect of the new ministerial leadership could be felt in November 1948. This was when significant changes took place in teaching tactics: these continued along the lines of the freshly translated Soviet military regulations. As a former student remembers, "Terrain tables made their way into the classrooms and we once again began to study tactics on squad and company levels."[30]

By then several former excellent teachers— among them Lieutenant General Sólyom, who was appointed as chief of the General Staff on December 3, 1948—were not teaching at the Academy. Just as a matter of curiosity, at this point in time the hourly teacher's fee was 25 HUF in the case of main subjects, while 15 HUF was paid hourly for subsidiary subjects.[31]

All this, however, did not create a problem for long, because— as it soon turned out—the end was around the corner. On January 6, 1949, Lieutenant General Sólyom as chief of General Staff informed the students that at the order of Minister of Defense Mihály Farkas the operation of the Military Academy was to be suspended so that the teaching staff would have the chance to acquire the new Soviet regulations. He also assured them that the second-year students would be considered as graduates of the Academy and would have higher positions secured for them at the General Staff or at the Ministry of Defense.[32]

Truly, according to one of the witnesses of the events, the educational activities were terminated on January 10, 1949.[33]

This document did not try to justify the suspension with the under-preparedness of the teaching staff but reasoned with "the necessity of restructuring the higher-level officer training." This was based on the December 1948 resolution of the Military Sub-committee of the Hungarian Workers' Party, which required that the General Staff should be terminated. The decree was issued on February 1, 1949.[34]

Lieutenant General Sólyom kept his promise, and placed about 21 students in high positions at different General Staff, Ministry of Defense and central organizations, in spite of the fact that the afore-mentioned decree of February 1 downgraded them to "intermediately qualified officers."[35]

With the suspension of the *Honvéd* Military Academy's education the first—extraordinarily important—section of developing a post-World War II higher officer-training program was closed.

Notes

1. Dr. Lajos Móricz, "A magasabb tisztképzés magyarországi történetének elvi, szervezeti és vezetési kérdései" [Theoretical, Organizational and Leadership Questions of the Hungarian History of Higher Officer Training], *Akadémiai Közlemények* 178 (1991): 9.
2. Dr. Ferenc Lengyel, *A magyar tisztképzés rövid története* [The Short History of Hungarian Officer Training] (Budapest, 2000), p. 31.
3. *Honvédségi Közlöny* (hereafter cited as HK) 39 (April 10, 1947) 12: 117.
4. This is also supported by the results of the entrance examinations that took place between September 15 and 20. According to the entrance examinations all the admitted 18 officers but one served as officers before the war and except for four all of them graduated from the Ludovika and Bolyai Military Academies. See Dr. Antal Oroszi, *A Zrínyi Miklós Katonai Akadémia történetének összefoglalása (1947–1996). A magyar katonai vezető- és tisztképzés története (Tanulmánygyűjtemény)* [Summary of the History of the Miklós Zrínyi Military Academy, 1947–1996. The History of the Hungarian Military Leadership and Officer Training (Collection of Studies)] (Budapest, 1996), p. 249.
5. Dr. Lajos Móricz, "A Honvéd Akadémia" [The *Honvéd* Academy], *Akadémiai Közlemények* 194 (1992): 28.
6. Dr. Lajos Móricz, *Békében és háborúban. Egy magyar katonatiszt emlékei a XX. század nyolc évtizedéről* [At Peace and at War: A Hungarian Officer's Memories of Eight Decades of the Twentieth Century] (Budapest, 2000), p. 156; Dr. Lajos Móricz, "Adalékok a Zrínyi Miklós Katonai Akadémia történetéhez" [Additional Material on the History of the Miklós Zrínyi Military Academy], *Honvédelem* 11 (1987): 76; Dr. Antal Oroszi, "Ötven évvel ezelőtt..." [Fifty Years Ago...], *Új Honvédségi Szemle* 12 (1997): 63.
7. The research of Dr. Antal Oroszi, colonel (ret.).
8. Móricz, "A Honvéd Akadémia," p. 29; Móricz, "Adalékok a Zrínyi Miklós Katonai Akadémia történetéhez," p. 77; Móricz, "A magasabb tisztképzés," p. 9; also see Dr. Antal Oroszi, *A Zrínyi Miklós Katonai Akadémia történetének összefoglalása (1947–1996)*, pp. 249–250.

12   Miklós M. Szabó

9. Honvédelmi Minisztérium Hadtörténelmi Levéltár [Ministry of Defense, Military History Archive] (hereafter cited as HM HL) 15,000/eln. szerv.–1947.
10. HM HL 24,746 sz./Eln.Kat. csf.–1947: the research of Dr. Antal Oroszi, colonel (ret.).
11. Oroszi, "Ötven évvel ezelőtt," p. 60.
12. László Csendes and Tibor Gellért, Kronológia a Honvédség történetéből 1945–1990 [Chronology of the History of the Honvédség, 1945–1990] (Budapest, 1993), p. 61.
13. HL Honvéd Kossuth Akadémia Parancsnokság [HL Honvéd Kossuth Academy Headquarters] (hereafter cited as HKA) 795/Eln.szü.– 1947. These headquarters had to notify the admitted students (Captains Bertalan Kandó and Lajos Móricz).
14. Móricz, "A Honvéd Akadémia," p. 29.
15. Oroszi, "Ötven évvel ezelőtt," p. 61.
16. HK 33/1947 (11.20), p. 307; HL HM 15,000/eln.szerv.–1947.
17. Dr. Lajos Móricz, "A Honvéd Hadiakadémia hallgatója voltam" [I Was a Student of the Honvéd Military Academy]. Honvédségi Szemle 8 (1983): 91.
18. László Sólyom, "Az új Hadiakadémia" [The New Military Academy], Honvéd 12 (1947): 29–33: the research of Dr. Antal Oroszi, colonel (ret.).
19. Móricz, "A Honvéd Hadiakadémia hallgatója voltam," p. 92.
20. Móricz, "Adalékok a Zrínyi Miklós Katonai Akadémia történetéhez," p. 77.
21. Móricz, Békében és háborúban, p. 156.
22. Móricz, "A Honvéd Akadémia," p. 30.
23. Móricz, "A Honvéd Hadiakadémia hallgatója voltam," p. 93.
24. Oroszi, A Zrínyi Miklós Katonai Akadémia történetének összefoglalása (1947–1996), p. 250.
25. Csendes and Gellért, Kronológia a Honvédség történetéből 1945–1990, p. 70.
26. HK 40 (April 1, 1948) 10: 121–122.
27. HK 40 (June 1, 1948) 15: 205.
28. Móricz, "Adalékok a Zrínyi Miklós Katonai Akadémia történetéhez," p. 78.
29. Oroszi, "Ötven évvel ezelőtt," p. 63.
30. Móricz, Békében és háborúban, p. 159.

31. General order of HK 1948/10 (April 1), 122. 2,700/eln.kv.–1948. "Compared with its value of August 1, 1946, the forint was 42% of its original value by the end of 1952, 33.5% of its original value by the end of 1960, 32.9% of its original value by the end of 1967, 23.3% of its original value by the end of 1978, 15.5% of its original value by the end of 1985, and 4% of its original value by 2000." Source: *Révai Új Lexikona*, vol. VII (p. 508, paragraph 2). It was published in 2001. In 1952 $1 was worth about 18.41 HUF.

32. Oroszi, *A Zrínyi Miklós Katonai Akadémia történetének össze-foglalása (1947–1996)*, p. 251.

33. Móricz, *Békében és háborúban*, p. 160; Móricz, "A Honvéd Akadémia," p. 31.

34. Dr. Gyula Balogh, *A hadiakadémiai képzéstől a szervezett doktori (PhD) képzésig. Tanulmány* [From the Military Academy Training to the Organized Doctoral (PhD) Training. A Study] (Budapest, 2001), p. 9.

35. Móricz, *Békében és háborúban*, p. 161; Balogh, *A hadiakadémiai képzéstől*.

Chapter II
The Impact of the Break between Stalin and Tito
on the Armed Forces of East Central European Countries

The facts mentioned in the previous chapter—among others, Mihály Farkas's appointment as minister of defense, and the dissolution of the *Honvéd* General Staff and the Military Academy— brought about a new era in the life of the Hungarian armed forces: the *Honvéd* Army was transformed into a People's Army (*Néphadsereg*). The essence of the change was that the Hungarian armed forces became a duplicate of the Soviet armed forces. Without this clarification the events occurring in the military higher education would not be understandable.

Since an armed force is a product of the political system, the cause of the aforementioned changes should be found in the changes of the political system.

The liberation from the Nazi occupation was followed by the unrestricted rule of the Allied Control Commission over Hungary. The Commission functioned until the Paris Peace Treaty came into existence on September 15, 1947, and, as a result, no government could be formed without the knowledge and the consent of its president, Marshal of the Soviet Union Kliment Yefremovich Voroshilov.

The absolute winner of the national elections of November 4, 1945, the Independent Small-holders', Agrarian and Citizens' Party, the main force for a thrust towards democracy, received 57 percent of the votes, while the Hungarian Communist Party got only 17 percent.[1] Democracy seemed to have won. Yet Voroshilov permitted the formation of the government only on the condition that the Hungarian Communist Party would receive the internal affairs portfolio, providing the minister of the interior, and the leadership of other important ministries. Simultaneously, the introduction of the Soviet-supported single-party system commenced in Hungary, and was completed in "the year of change," 1947–1948. Yet the country remained occupied by the Soviet army until the Austrian state treaty was concluded in 1955, since the roads linking the Soviet Union with Austria were secured by Soviet troops in Hungary. The Soviets were not supposed to interfere in Hungarian internal affairs, yet they did.

Right before the conclusion of the Austrian state treaty the Warsaw Pact[2] had been formed; this organization—not legally but in effect—subordinated the captive nations to the Soviet Union. This entailed a continued occupation and interference with internal affairs.

The Soviet interference in Hungarian affairs—including the fate of the armed forces—imposed two contradictory developmental processes on the country. Until 1948 the slogan was spread that "the army of the glorious Soviet Union will protect the country, and there is no need for a Hungarian army." The *Honvédség* was so decimated that it became the shadow of its previous self. However, on June 28, 1948, Tito was excluded from the Cominform, and thereafter the tide turned in East Central Europe. Stalin decided that the Tito regime had to be eliminated at all costs.[3] Correctly assessing historical patterns, Stalin expected Yugoslavia to put up armed resistance to any Soviet invasion. Yet he knowingly accepted war as a means to his political goal, undaunted by the dubious reputation that he would thereby have acquired as the first to initiate a war between socialist states.[4]

From the summer of 1948 onwards, the armies (in being or to be created) of East Central Europe figured as active participants in Stalin's military designs against Tito. To realize his goal, the earnest and concentrated effort of which totalitarian states are capable was made to prepare the captive socialist states rapidly for war. What happened in Hungary demonstrated the character of the process in all of East Central Europe, just as a drop of sea water reflects the chemical composition of the ocean.

A transformation started in the status of the Hungarian armed forces, from national army—*Honvédség*—into "people's army," from a ghost into a modern fighting force. Farkas rapidly took the reins in hand, and things began to happen in the army.

The main innovations were the following: constant anti-Titoist indoctrination; the introduction of Soviet "advisers"; the reorganization of the supreme command; the establishment of the Soviet *politruk* (political officer) system; the stabilization of the Party hierarchy and Party cells in the army; re-equipment with Soviet military hardware; the introduction of Soviet regulations into all areas of military life; the intensification and expansion of new officer training courses; an increase of the army's strength; periodic purges; the reorientation of the Hungarian production system into a war-industrial complex; the integration of this new People's Army into Soviet military operational plans against Yugoslavia. In short, what was begun on September 29, 1948, was a protracted process of Sovietization of the Hungarian armed forces, just as the armed forces of all the other East Central European socialist states were being Sovietized, for incorporation into the anti-Tito military machine.

The anti-Titoist indoctrination and propaganda were intended to provide a *raison d'être* for the revival of the army. Titoist Yugoslavia was depicted as a lethal threat to Hungary. The Soviet "advisers" tolerated no deviation either from the incessant anti-Titoist indoctrination or from an absolute concentration on a single problem: the coming war against Yugoslavia.

A few weeks after the arrival of the "advisers," blueprints for the new supreme command had been translated into Hungarian. The minister of defense became not only the political head of the department, but also the commander-in-chief. The position of first deputy minister of defense was created. He was the head of a huge new Chief Section for Political Affairs. To fill this position, a Party

Central Committee member, Sándor Nógrádi, like Farkas a former Muscovite émigré, was appointed. This new section had a double function. It was both a major department of the Ministry of Defense and a central branch of the Secretariat of the Central Committee of the Hungarian Workers' Party (Communist Party). In addition to his political and educational mission, in a purely military sense the first deputy minister wielded immense power over the army. This power derived from the fact that his direct subordinates, the *politruks*, were co-commanders of all military units at every level down to battalion headquarters. Furthermore, the Party cells were not autonomous organs within the army; they were directly subordinate to the *politruks* of their units. The network of officials reporting to the first deputy minister penetrated all units. Party trusties, subordinates of the *politruks*, some publicly named, others secret agents, watched every armed man, Party member or not, and reported to the *politruks*.

Under the supervision of Soviet "advisers," indoctrinated by the *politruks*, and watched around the clock by the secret police, the Hungarian army grew rapidly. In September 1948, the Hungarian *Honvédség* had consisted of two undermanned, under-equipped infantry divisions and one engineering division. No armored force or air force had existed to speak of. The *Honvédség* had been deliberately left to disintegrate, and amounted to nothing that would warrant being called an army. Once the supreme command was reorganized, the spectacular growth of the new People's Army started. By the summer of 1950 it consisted of nine infantry divisions, two armored (mechanized) divisions, four artillery brigades, nine anti-aircraft divisions, three engineering brigades, one chemical battalion, one cavalry brigade, one signals regiment, one communication brigade and three heavy armored regiments.

The total strength of these forces amounted to 250,000 men (about three-and-a-half times the number permitted by the Paris Peace Treaty). The air force consisted of one air fighter division.

The feverish modernization of airfields and the speedy construction of numerous new ones were undertaken, so that the Soviet air force could also use Hungarian air bases against Yugoslavia.

In addition to these forces, there were the two squadrons of the Danube Fleet.

In all, it was by far the largest force that Hungary, including even the old Hungarian Kingdom, had ever kept under arms in peacetime. Taking into account Hungary's limited national resources, it was an enormous armed force for a small country to support in wartime, let alone peacetime.

The last year of the 1940s and the early 1950s were a period when life in Hungary was completely geared to the coming war: politically, economically, militarily and psychologically. It is no exaggeration to say that war in the early 1950s was imminent. The war that the USSR was preparing against Yugoslavia was to be a war of aggression, of course, violating international law and the United Nations Charter. In fact, by forcing Romania, Hungary and Bulgaria to keep such huge armed forces in readiness, the Soviet Union had violated international law even without going to war. All three countries, losers in the Second World War, had signed the Paris Peace Treaty, which restricted their armed forces. All three countries were now compelled to maintain armed forces between two and three-and-a-half times as large as they were entitled to by the treaty.

The rapid and significant deterioration of the standard of living of the Hungarian population between 1950 and 1953 was mostly due to the investment in heavy industry and military manufacturing. Neither was in proportion to the country's economic capacities.

In 1950 two major events occurred: a massive purge of the Hungarian strategic leadership and the outbreak of the Korean War. They both had a cause-and-effect relationship with what transpired.

The generals and colonels of the Hungarian People's Army were ordered to appear in the main hall of the Officers' Club at noon on May 23, 1950. No reason was stated. Minister of Defense Farkas was the main speaker. He stated that the most vicious conspiracy had been uncovered in the highest echelons of the People's Army and the spies apprehended. The arrested men were the following: Lieutenant Generals László Sólyom, chief of the General Staff, and Gusztáv Illy, head of the Personnel Chief Section; Major Generals István Beleznay, commander of the army corps of Székesfehérvár, György Pórffy, commander of the artillery, Kálmán Révay, commander of the mechanized troops, and Dr. Merényi-Scholtz, surgeon-general of the army; General Staff Colonel Sándor Lőrincz, head of the Financial Chief Section of the General Staff. All of them were charged with being Tito's stooges, and hence enemies of the great protector Stalin. They were all executed shortly thereafter.

Thus all those high-ranking officers who had started their careers in the *Honvédség* (Royal Hungarian National Army, or simply Hungarian National Army after a republic was proclaimed in 1946) were eliminated from the strategic leadership of the People's Army, and were replaced by "worker cadres," half-trained persons who were considered to have been loyal to Bolshevism. The real strategic leadership thus fell into the hands of the Soviet "advisers." The Hungarian army was ready to participate in Stalin's war on Yugoslavia.

It is a sad irony that, shortly before the outbreak of the Hungarian Revolution in mid-October 1956, all of these innocent victims of the Stalinist purge were reburied with full military honors.

On June 25, 1950, the alarming news was broadcast that North Korea had launched an offensive against South Korea. The immediate result was not a propaganda campaign against South Korea, but an unbridled hate campaign against Titoist Yugoslavia. North Korea launched the war exactly when the Soviet bloc in Europe was ready to initiate the aggression against Yugoslavia.

That coordination indicated that there was a direct relationship between the timing of the Korean aggression and the completion of preparations for war against Yugoslavia.

Planning the expansion of the Communist-controlled world in Asia and in Europe almost simultaneously was based on the Chinese leader Mao Zedong's view that "America was a paper tiger" that would never interfere with armed forces if the Communist world started expansion by force.

Stalin hoped that Mao's claim would prove correct. In that case the next victim of Communist aggression would have been Yugoslavia. That was the reason why both these wars were prepared so that both could be started simultaneously.

On the other hand, Stalin was unwilling ever to start a war in which the Soviet Union would have faced the USA as the opponent. Mao proved to be wrong: the USA and the West were not "paper tigers"; instead, they defended the independence of South Korea, and so saved Yugoslavia also.

As long as Stalin was alive he never openly stopped the anti-Tito propaganda, but he decided not to attack Tito's Yugoslavia, so as to avoid possible armed confrontation with the USA. Only after his death, in 1953, was this decision made open.

The Soviet and East Central European captive nations' armed forces were gradually reduced; thousands of their officers were dismissed from the reduced armed forces. In 1956 this reduction of the Hungarian army was still in progress; a disappointed officer corps faced the freedom fighters.

That is the political as well as the general military framework within which we should study and understand the development of the system of higher military education in Hungary between 1947 and 1956.

Notes

1. The results of the general elections held on November 4, 1945, in *Magyarország történelmi kronológiája* [The Historical Chronology of Hungary], vol. IV: *1944–1970* (Budapest, 1982), p. 1023.
2. René Albrecht-Carrié, *A Diplomatic History of Europe since the Congress of Vienna* (New York, 1958), pp. 411, 644.
3. Cominform: the Information Bureau of Communist and Workers' Parties between 1947 and 1956.
4. See Béla K. Király, Barbara Lotze and Nándor F. Dreisziger, eds., *The First War between Socialist States: The Hungarian Revolution of 1956 and Its Impact* (New York, 1984), pp. 3–31, and Wayne S. Vucinich, ed., *At the Brink of War and Peace: The Tito–Stalin Split in Historic Perspective* (New York, 1982), pp. 273–288.

Chapter III

The History of the *Honvéd* Staff Course
(July 22, 1949–September 25, 1950)

The highest-level military educational institution in Hungary, the *Honvéd* Military Academy—established during the fall of 1947—ceased *de facto* to exist only by the suspension of its operation on January 17, 1949,[1] since Minister of Defense Mihály Farkas—promoted to colonel general[2] on April 22, 1949—moved Lieutenant General István Kozma of the *Honvéd* General Staff to the Military Academy, effective from June 1, and commissioned him to attend to its commander's tasks.[3] In an earlier letter[4] Farkas had transferred Medical Corps Lieutenant Colonel Dr. Gyula Szeremley as political officer to the Academy—in accordance with the *Honvéd* Political Officers' Institution established on February 19, 1949, and with the "co-commander" system.[5] Hence, for several years, there were two signatures on every document: those of the commander and the political officer.

Yet the minister's order was also a retrospective sanctioning of the system, since on May 17, 1949, Lieutenant General Kozma asked Colonel General Mihály Farkas to ensure personnel needs based on the organizational and staff chart of the Academy approved on May 9.[6]

As far as the established ranks of people of significant status were concerned, the following were the guidelines: the commander was to be a lieutenant general, the chief of staff was to be a major general or a colonel, the head of educational activities, heads of departments and the head of Logistical Services were to be colonels, the department head of training and the head of the Students' Section were to be colonels or lieutenant colonels, and the teachers and the head medical doctor were to hold the rank of colonel or major.

22

Major General Béla Kerekes   Lieutenant István Kozma
Inventory number: 2198   Inventory number: 82566

The political officers assigned to the commander and to the head of the Political Department were colonels, those assigned to the chief of staff, to the head of educational activities, to the commanders of the General Military Department and military services and the lecturers of the Political Department were captains or majors, and those assigned to the ten departments held the rank of lieutenant or captain.[7]

"The appointment of Lieutenant General Kozma was primarily urged by Lieutenant Pálffy. His arrest and execution also sealed the commander of the Academy's fate. (He was forced into early retirement in August 1949 and was later executed for war crimes.)"[8]

By the early fall of 1949 the name of the institution had been changed to "Staff Course." It had been settled at the Hungária körút base on the site of today's Miklós Zrínyi University of National Defense.[9] A situation report of September 15[10] gave an account of the preparation of the Staff Course in a rather critical way. It claimed that the rather low numbers of assistant personnel would make the training of the students difficult, and considered the available regulations, assistance and instructions to be insufficient, while there were hardly any people suitable to evaluate the lectures. Yet launching the academic year on October 17, 1949, was considered to be realistic.

Accordingly, the relevant parties made arrangements for the distribution and allotment of the students of the Staff Course. The planned 110 people included 10 infantrymen, 25 artillerymen, 11 anti-aircraft artillerymen, 10 air force men, 2 sailors of the river fleet, 6 men of the engineering corps, 6 men of the signals service and 10 men of the armored troops. Some of the students were supposed to participate in higher commander training and higher staff training, while others would receive regimental commander or regimental staff training.[11]

The Staff Course—subordinated to the chiefs of General Staff—was organized according to the decree effective from October 1, 1949. The essence of its constitution was that the chief of staff and Logistics Department were subordinated to the commander of the Course, while the Political Department and the library were subordinated to the political officer. Major General Béla Kerekes—a former aide-de-camp to President of the Republic Zoltán Tildy—became the commander of the Staff Course.[12]

The Ministry of Defense assigned 12 political officers to the institution on October 10, 1949. Lieutenant Colonel Dr. Gyula Szeremley became the political officer of the Staff Course, and Lieutenant Sándor Stráhl became the political officer of the chief of staff of the Course.[13]

During these days Lieutenant General István Kozma was arrested, and Lieutenant General György Pálffy was executed on October 24.[14]

From these days onwards—due to the war preparations against Tito—even educational institutions received code numbers. In this way the *Honvéd* Staff Course became Honv. 10. PF. 8985.[15]

The Telephone Factory, a joint-stock company, donated a flag to the Staff Course.[16] Lieutenant General Sándor Nógrádi, the deputy minister of defense, organized celebrations in the Telephone Factory on December 19 and at the Course on December 20—one day before Stalin's birthday.

In December 1949 the *Honvéd* chief of General Staff submitted a "Situation Report" to the minister of defense.[17] According to the report, "the Stalin-initiated work competition had a very favorable effect on elevating our students' knowledge." In this way the monthly grade point average (GPA) of the Course increased to 4.0 from 3.66, while in the case of several students this entailed a more dramatic increase, from 2.8 or 3.5 to 5.0. However, this could not solve all the problems. As is spelled out by the situation report, although the curriculum had been revised—and as a result, it became transmittable and understandable—the acquired knowledge did not become lasting, because there was little time dedicated to practice. The conclusion was that more time should be dedicated to individual learning. Since a packed curriculum was the order of the day in any case, this could be done only by reducing the length of classes to 45 minutes. Accordingly, the students' daily schedule was as follows:

8:00–13:20 six 45-minute classes with 10-minute breaks between classes;

13:20–15:20 lunch;

15:20–17:00 two 45-minute classes;

17:15–19:45 individual studies;

19:45–20:30 dinner;

20:30–22:30 individual studies.

Major General Béla Kerekes, head of the Course, and Lieutenant Colonel Gyula Szeremley, its political officer, submitted a report on January 5, 1950, to the chief of General Staff about the organization of the Staff Course.[18] The complicated nature of the previous months' activities is clear from this report, since the decree on the foundation of the Staff Course had to be modified and/or complemented five times during five months.

The chief of staff and the Logistics Department commander, as well as the Human Resources Sub-department, were directly subordinated to the commander of the Academy. The "D" (Security) Department was subordinated partly to the commander and partly to his chief of staff. Interestingly enough, however, the Party secretary of the Staff Course was accountable *exclusively* to his/her supervisory Party organization. The "Academy political officer" was the head of the Department of Political Studies and the club and library.

The (regular/effective) number of the Staff Course during this period of time was 57/63, with the following breakdown: 13/13 political officers, 13/14 subalterns, 3/3 non-commissioned officers, 19/19 regular officers, 64/60 *honvéd*s, 116/116 officer students and 82/82 civilian employees, that is, 354/357 altogether.[19] Consequently, there was a minimal shortage among the enlisted, while in several places there were some redundant personnel.

The GPA of January 1950 decreased to 3.8, which Major General Kerekes and staff explained by the disruption caused by the Christmas break. Yet they considered the switch over to 45-minute classes to be a success. Once again they complained about the low number of teaching staff, but happily reported that Major General Béla Király, commander of the *Honvéd* Infantry, and Major General György Pórffy, commander of the *Honvéd* Artillery Troops, visited the Academy during the month.[20]

In February 1950 Major General Imre Somogyi, head of the Logistics Department, made an attempt to categorize the subjects—

by differentiating between major and minor subjects—in order to calculate teaching bonuses. As a result, during the teaching of the students of all the branches the following subjects qualified as major ones: political studies (history, political economy and questions of Marxism-Leninism), general tactics, the methodology of tactical training, army organization, military history, infantry shooting and arms studies, topography, Russian language and mobilization. The following received minor subject classification: logistical service, physical education, basic training, Army Regulations, issues of military traffic and branches of military service studies. Besides these, there were specially dedicated major and minor subjects for the students of different military branches: artillery, anti-aircraft artillery, armored troops, engineers, signals troops, air force and river flotilla.[21]

Regarding the further training of officers, the chief of General Staff held an important conference on February 27, 1950, where the officer-training plans of 1950–1951 were announced. The essence of the Plan was the following:[22]

1) The foundation of a *tactical leadership-training* academy served as a basis for the further development of the present Staff Course, where suitable officers of different branches were trained to become tactical and operational leaders (commanders of regiments, divisions, army corps and their staffs, as well as higher commanders of different branches).

2) The present branch schools were to be further developed to educate squad, sub-unit, unit and regimental commanders, as well as divisional branch commanders.

Accordingly, in the fall of 1950 two kinds of training courses within the *Honvéd* Staff Course were supposed to be launched: a one-year course and a two-year course:

One-year course: Considering the need of the fall of 1951, and the number of people whom we have presently and who

are graduating from special courses and from the Staff Course
until then, we will train the necessary contingency of commanders specified above for the fall of 1951.

Two-year course: Considering the natural dropout rate
expected by the fall of 1952 and the expected need of central
organizations, we will train the extra demand demonstrated
between the fall of 1951 and 1952 during a two-year course.[23]

As far as the "expected natural dropout rate" was concerned, it
is debatable whether it was wise foresight or whether the plans for
the purges had already been completed, since between January 15
and June 19, 1950, 9 generals, 115 high-ranking officers and 110
subalterns, 234 people altogether, were removed from the army.
This was, however, only "the beginning," since 10,000 more people followed suit during the second term. Preparations for a
Yugoslav war also played a role in this, besides the political purges,
and as a result the effective force of the army was on course for a
rapid increase in numbers: it was a 30,000-strong force in 1949,
while the "Bem" organization plan envisaged a military force of
45,000, which, by the end of the year, had gone up to 114,896,
including 13,361 officers.[24]

Returning to the subject matter of the conference of February
27, it spelled it out that, although the students were "theoretically"
accepted from the different branches, "as the Staff Course most of
all caters for general tactical and operational knowledge, it would
also be possible that qualified officers of different branches would
get into higher commanding positions (divisional or army corps
commanders) after completing the Course. Therefore, during the
entrance examination process no distinction was made between the
different branches. "Those were admitted who were suitable, voluntarily applied for higher military training, and passed the
entrance examination."[25] Partly based on these principles, partly
based on some calculations, 60 people from the infantry branch

were to be sent to the one-year and 45 people were to be sent to the two-year course.[26]

For political reasons, on March 6, 1950, a statutory decree deleted the "*Honvéd*" adjective from every military institution, and therefore from the *Honvéd* Staff Course as well.[27]

In March 1950 the leadership of the Staff Course tried to develop a uniform evaluation system of the students,[28] which was supposed to be implemented by the end of the first term. The main goal of the system was that the evaluation of the most important military subjects would be emphasized when considering the students' academic advancement, and therefore *grading index numbers* were introduced. Prior to that, the 1 through 7 grading system was changed to 1 through 5 and the requirements connected to the different grades were also defined at each evaluation. The grades and the requirements were as follows:

"5," that is, an "outstanding" grade, went to a student who had acquired knowledge of the material perfectly and was able to apply it in practice "without any gaps," whose attention covered every detail and who organized cooperation successfully, who solved his tasks well *without his principal's or teacher's intervention*, who stood out among his fellow students with regard to his aptitude as a commander, and who emphasized self-education.

"4," that is, a "very good" grade, went to a student who had satisfied all the aforementioned requirements, except for the fact that he solved his tasks *with his principal's or teacher's minor intervention*, and who was "above average" among his fellow students with regard to his aptitude as a commander.

"3," that is, a "good" grade, went to a student who had acquired knowledge of the study material well and who presented that "briefly, accurately, coherently and in precise military terms; however, his responses were sometimes inaccurate, but they did not harm the essence of the question"; he solved his tasks *without major help* and usually well.

"2," that is, a "satisfactory" grade, went to a student who had "necessarily acquired knowledge of the main material of the training plan, which secured further development for him; he presented his knowledge quite aptly and tactically lucidly, but there were inaccuracies in his responses that revealed that he did not know the details of the subject"; he was able to solve his tasks only with *significant help.*

"1," that is, a "not satisfactory" grade, would go to a student who had only superficial and ungrounded knowledge about the *essence* of the question, who could not apply his theoretical knowledge in practice, and who could not solve his tasks even with substantial *help.*

Besides these ground rules and beyond knowledge, the teachers also had to consider such factors as practicality and verbal and written articulation when grading the students.

This was followed by establishing *an individual ranking order.* So-called index numbers were established, which were the following, for example, in the case of a field artillery student: *3x* for political economy, political history, Marxism-Leninism, general tactics, staff service and field artillery tactics; *2x* for field artillery shooting training, logistical service, topography and infantry shooting training; *1x* for every other subject.

The grade received in a given subject had to be multiplied by the relevant index number of that subject, and the values received had to be added up and divided by the sum of the index numbers. In this way, an average value ranging between 1 and 5 was the result, which had to be calculated to two decimal points. The *students' ranking order* was established by the *average value of the academic sum results* received by means of this method.

In sum:

"Outstanding" was given to the students who—according to the calculation above—achieved or surpassed a 4.8 GPA and did not have a grade 4 (or weaker results) in general tactics or in political subjects (Marxism-Leninism, history and political economy).

"Very good" was given to the students who achieved at least a 3.7 GPA and did not have a grade weaker than 4 in the aforementioned subjects.

The final evaluation of "good" was given to the students who achieved or surpassed a 2.6 GPA.

A condition for belonging to any of the three aforementioned categories was the fact that the student would not have received a single grade 1.

A student's achievement was evaluated as "satisfactory" if he achieved a 1.5 GPA, while the ones with a GPA below 1.5 received a "not satisfactory" qualification.

Strengthening the squad spirit was considered to be an important element of community education, and therefore squads were ranked according to the following model: the grade point averages of the students were added up and divided by the number of students. The value gained in this way constituted the basis for evaluating the squads and comparing their academic results. Accordingly, the squad whose GPA—calculated by the above steps—was at least 4 became "outstanding"; with a 3.2 or higher GPA it was "very good"; with a 2.5 GPA it was "good"; with a 1.5 GPA it was "satisfactory"; if it did not achieve a 1.5 GPA, it was qualified as "bad" (not satisfactory). Moreover, in the case of the first three qualifications the number of grades 1 or 2 could not surpass 20 percent of the total number of students.

It is difficult to fathom why certain people could not hold their tongues in this appalling period, laden with show trials and showdowns. According to one of the documents, the following took place:

On January 23, 1950, Corporal Elemér Varga read headlines and notices in bold type from the newspaper in the room for rank-and-file soldiers. When he read "the beloved leader of the Hungarian people, Mátyás Rákosi, appeared at the celebration," Varga added "Yeah, I do love him, don't I? Like a goat likes a knife." Prior to this incident... he called Comrade

Stalin an "ancient post office robber"... On February 21, 1950, Lance Corporal János Szilágyi prepared a drawing and Corporal Varga commented, "Why on earth are you destroying this picture with that political officer? You have quite defaced it! Why do you place the flag in the hand of the chief sergeant? He won't be in the front in a battle, for Heaven's sake! When we retreat, at the most!" On March 9, 1950, after an *ad hoc* political meeting relating to Mátyás Rákosi's birthday, in the room for rank-and-file soldiers, Varga asked Honvéd János Mihályi "How old is Rákosi?" and after the response he said "I never knew that shit lasts that long!" In mid-January 1950—while decorating the site—Varga made the following remark about Comrade Lenin: "How much we have to work for the sake of this old fart!"... when one of his fellow soldiers cheered Comrade Stalin on seeing his illuminated photograph, Corporal Varga made the following remark: "As far as I am concerned, he can live long but only on this food."[29]

The investigators of the case considered Varga's working-class background to be an aggravating circumstance in his demonstrating this destructive behavior, and due to the fact that he was of good intellectual abilities it could not be assumed that he would not have made his remarks with destructive intentions. As a result of all the above, on March 24, 1950, at 8:00 p.m., Corporal Varga was taken into custody and criminal proceedings were initiated against him.

There were those who tried to stand out differently. Lieutenant Pál Kurilla wrote a paper on Marshal Suvorov. The paper itself is not known, but the letter written to the author was signed by Lieutenant Colonel Ákos Vizi-Makkay, chief of staff, and Captain Sándor Stráhl, political officer. Kurilla was praised for his enterprising initiative and for his remarkable ambition in that he wanted to publicize the life of a Russian military leader, but in its present form it was deemed not suitable for publication in a military journal. The reviewers pointed out that the dominant theme of the essay was not correct, because it discusses Suvorov as a great military leader. It was not pointed out that he gained huge distinction in solidifying the Russian Tsarist Empire; he was a strategic master... but he had to utilize both his military art and his abilities as a military commander for the consolidation of Tsarist rule and for breaking down European progressive freedom movements. Therefore, the military commander of the Tsar and the military commander of soldiers need to be separated. He primarily achieved the position that he occupies in present-day Soviet military history as the commander of his soldiers.[30]

(Furthermore, even though the last paragraph was encouraging—"Revising your presentation according to the above remarks and in their spirit, it will, by all means, become publishable"—I could not find any trace of Lieutenant Kurilla's further career as a writer.)

The situation report of the month of March[31] demonstrated the problems raised by preparations for the April 4 military parade: classes of military history, army organization and rules of military service were cancelled, and there were other—replaceable—cancellations. Yet the exam result of the term completed in March was a 4.0 GPA, mostly because the previous weak performers had developed greatly.

In the meantime, work regarding the improvement of education continued. Accordingly, Minister of Defense Colonel General Mihály Farkas established the system of examination procedures and grading in his document issued on April 22, 1950.[32] Its essence concerning the educational institutions was the following:

- It introduced the 1 through 5 grading system.

- Final examinations had to be taken in every educational institution and in every course of the following subjects: political, formational and sport training; rules of the given military branch; "General principles for the officers of the *Honvédség*" (in future use in documents the latter was referred to as "Technical minimum" or "Tech minimum"); five or six main subjects of the branch represented by the Course.

- Only those could be submitted to a final exam who demonstrated at least a "satisfactory" result during the academic year;

- The final exam had to be taken before a sub-committee after drawing an item and after a preparation time of 15–20 minutes. Each member of the sub-committee carried out his/her own evaluation.

Commander of the Further Training Course Major General Béla Kerekes and Political Officer Major Zoltán Horváth submitted a report on April 26, 1950,[33] to the chief of General Staff on the experiences of the first term of the Staff Course. The essence of the submission was the following:

- At the beginning, the lectures were too lengthy and pitched at too high a level.

- The basic moral of the story was that only the most essential and most important materials, and only subjects that were indispensable for military training due to their fundamental significance, should be taught in the future.

- The desire to learn, diligence and willpower were at a maximal level.

- The GPA increased from 3.6 to 4.

- Major differences posed substantial problems: the 22–44 age range; education, which ranged from the first three primary school classes to a college degree; military service, which comprised a range between five months of military service and a several-year-long army career; family status, which ranged from unmarried to fathers with five children, and so on.

- Originally, 71.55 percent of the students were industrial workers, 12.07 percent were agricultural workers, and 16.38 percent were intellectuals.

- Originally, a weekly one-day (eight-hour-long) further training session was organized for the officer and subaltern staff personnel of the Course in the first few months, which was a major burden for them; therefore it was later transformed into a biweekly three-hour-long further training course for the teachers, which proved to be much more effective. Further training was complemented by individual studies; in this way, the fulfillment of all the requirements prescribed by the minister of defense was achieved.

- The Command tried to ensure that all the lectures would be completed by the beginning of the term, but this target was only 80 percent fulfilled during the ten weeks at their disposal.

In the meantime, a situation report was submitted on the April activities of the Staff Course[34] in which it was pointed out that definitely more time should be spent on the military activities of higher units. It was also becoming obvious that although military exercises involving two to three days of on-site activities proved to be extremely valuable, they also caused problems, because individual learning activities on other days decreased, and other subjects were crammed together. There were suggestions made to abandon the following subjects: the rules of military service, shooting training, management training, arithmetic, geometry and geography. Only the abandonment of the first two subjects was justified: studying the rules of military service would only be a review in any case, and only the theoretical part of the shooting training would be dropped. These ambitions were not very well supported by the results of the monthly shooting practices, where, for example, at the second pistol shooting training session only 20 out of 107 people produced acceptable results. As is evident from the above, the ideal of a highly educated officer corps was sacrificed on the altar of temporary problems—as became common practice during the following years and decades.

In the meantime, the esteem of the Staff Course within the army took a negative turn, since Major General Károly Janza, head of the Logistics Department,[35] informed the leadership of the Course that the salaries for commanders and political officers of the Course were to decrease to category 5 from category 3, but Major General Kerekes and Lieutenant Colonel Dr. Szeremley, MD, would continue to receive their former salaries as long as they stayed in their positions.

The situation report of May[36] raised several problems in a spectacular way:

- Dealing with higher units continued to remain a major problem;

- Academic discipline slightly declined and the better ones became over-confident;

- At the request of the students, time spent on individual studies increased, but, as the crushing authoritative opinion says, "a few of the less conscious comrades do not, for example, spend free afternoons studying but wish to utilize them for individual interests and objectives";

- The results of the repeated pistol shooting practice again became insufficient;

- In the interests of completing their lectures on time, the monthly teacher further training conferences were terminated (except for the political and Russian-language classes), yet teachers had to participate in the branch commandership further training sessions;

- Work continued to be hindered by the shortage of personnel.

The end result of the activities in June[37] became more positive, since studying of the whole training material was completed, the branch-related discussions on applications—included in the discussions on tactical applications—and the executed leadership practices bred very good results, and therefore more leadership practices were included in the following training year.

At the beginning of the month there was some setback on the academic front, since "some distrust surfaced among the students regarding the correctness and professional suitability of the material as a reaction to the revelations about the 'gang of spies' within the army (that is, the launching of show trials against Lieutenant Generals László Sólyom and Gusztáv Illy, Major Generals Kálmán Révay, István Beleznay, György Pórffy and Gusztáv Merényi, and Colonel Sándor Lőrincz)."[38] Since student soldiers were just like any other students in the 1950s as well, they grasped the opportunity and were ready to "fabricate ideologies": accordingly, "There were statements that it was not worth studying, because who knows whether the material presented to them is correct."[39] This slightly contradicted the teachers' opinion that the majority of the students had already achieved the level required for them to apply the acquired material in practice. During the report period more than half of the students produced "very good" results and only one of them received a "satisfactory" grade.

On July 21, 1950, the leadership of the Staff Course issued an extensive set of instructions in the domain of the training plan of the 1950–1951 academic year.[40] Hereby it was established that from the following fall onwards a one-year course and a two-year course would be held. The leadership of the Course recommended to the minister of defense that only officers with at least a one-year appointment giving department or higher command experience could participate in both training forms. This was the prerequisite for organizing troop training sessions and for higher units after completing the Course both in peacetime and in wartime.

The "most perfect iron discipline" was to continue to be solidified in the new year in the areas of internal order, internal service, material maintenance, cleanliness, clothing and behavior, and, first of all, in the area of carrying out orders. The academic discipline was to be elevated to an even higher level, so that after finishing their studies, the students would become firm and disciplined commanders of the *Néphadsereg* (People's Army), who would educate their troops in the same spirit. The tactical sensibility had to be developed and "the students have to acquire the Soviet tactical and strategic principles, so that in the possession of adequate objective knowledge, they would be able to apply our new and modern military principles appropriately, rightly, and correctly in practice as well."[41]

The Objective of the Training

One of the basic tasks of the training was spelled out in the following way: "We must train honest, just, brave, unselfish and responsible commanders who are ready to give their lives for their people, and who can train the sons of the people to become convinced fighters who are ready to fight for their country and can be led to victory."[42]

The following had to be acquired by the students during the training:

- The theoretical foundation of Marxism-Leninism;
- The bases of modern fighting, all the abilities needed for well-grounded decision-making and for organizing the successful cooperation of branches;
- The methodological regulations of shooting training and perfect firing;
- The principles of the training and educational methods of the *Honvédség;*
- The right practical application of principles of regulations and instructions.

As far as the essence of the July 1950 activity is concerned,[43] the training tasks were fulfilled, the academic progress of the students "could be considered to be exceptionally good," and the principles were well implemented in practice.

As it turns out from the situation report, on July 24 a Ministry of Defense Review Committee paid a visit and urged that the objectives, requirements and numbers of the forthcoming course would be specified. The situation report of August[44] already reflected the "end-of-year burst of speed." According to this, the two-day-long leadership exercise was successful; the Course closing exam was postponed from August 4–14 to September at the order of the minister of defense—who possibly wanted to play it safe and wanted to avoid any risk. Therefore, the free time gained could be spent on exam preparations, and on August 14 "trial exams" were held. Between August 16 and 31 a tactical trip was made.

Yet the state examination period was held at that time instead of the originally proposed period of September 2–8.[45] Therefore, the Course closing festivities were planned to be held on September 25, with 350 participants, principals, delegations, and the representatives of the Telephone Factory, which had donated the unit flag.

The September situation report[46] emphasized the fact that the final exams had been completed in order during the time period of September 2–8, most of the teachers and the students participated in a military exercise between September 9 and 24, and the end-of-year celebration took place on September 25 in the presence of the chief of General Staff. Thus the Staff Course as an independent institution ceased to exist, and parts of it were merged with the *Honvéd* Academy established under the leadership of Major General Béla Király on September 25.

Fortunately, a comprehensive and valuable material of 25 pages had been completed, entitled "Summary Evaluation and Situation Report on the Work of the *Honvéd* Staff Course: July 22, 1949, and September 15, 1950,"[47] and submitted by Major General Béla Kerekes, head of the Course, and Captain Béla Kránitz, its political officer, to the minister of defense.

This report gives a good overview of the one-year-long activities of the Course, which will be summarized below.

1. Organization and Preparation

The Staff Course was set up on July 22, 1949, and "on this day it began to prepare for a totally new style of higher-level officer training school based on experiences gained by visiting Soviet military schools."[48] The faculty comprised officers of excellent background who had formerly served in the Hungarian Royal *Honvédség* and who had subsequently been unemployed for an extended period of time. Therefore, "they had very little notion of modern Soviet training principles. Their knowledge was gained at and limited by our preparatory course held in the months of May through July." Furthermore, basic regulations were also missing, or very few were at their disposal, in rough translation. Similarly, it was of grave consequence that they did not dispose over any military equipment, blackboards or training tools.

For the sake of more effective preparation, mandatory accommodation in on-site housing was implemented for the participants, which did not kindle undivided enthusiasm among the parties concerned. However, the commander wanted to ensure that "the available short time would be effectively utilized by the teachers by saving on commuting time."[49] It was further justified by declaring that the word-for-word note-taking of the teachers' lectures could only be done in this way.

The Course was originally planned to last for six months, but it soon became evident—mostly due to the students' low level of preparation—that it had to be increased to one year. In retrospect the Course leadership self-critically stated that "we, however, committed the mistake of setting up the whole first term for preparations and, as a result, we only had half a year to realize all the original objectives set for the Course."[50] They also realized that too much time was spent on section, squad and company training. Yet even at the end of the year quite a few students proved to be quite unacquainted with section and squad leadership.

2. The Training of Students

The extreme heterogeneity of students raised serious difficulties during training. As was mentioned before, there were people of diverse background among the students, with the following demographics: age ranging from 22 to 44; education ranging from the first three primary school classes to military college or university degrees; marital status ranging from unmarried to married with five children; physical workers (71.55 percent), agricultural workers (12.07 percent) and intellectual workers (16.38 percent); people with a several-month-long Party school background, people with several-year-long Party work experience, and people with basic political knowledge. The writers of the report pointed the following:

> It is obvious that this vast diversity within the student body placed a heavy burden on both the teachers and the students. Common ground had to be found for the teaching material, for sensible and uniform requirements, and for practical evaluation system... The students' desire to learn, their willpower and their diligence could handle the sharpest criticism... It has been observed, though, that the students did not always recognize the most important questions... However, the good political work, the support of the Party and the directing work of the teachers always brought about decisive changes on this field as well... and a gradual and constant increase in the academic achievements took place.[51]

For the sake of historical authenticity, the above should be slightly modified. In terms of academic achievements, we witness a slight fluctuation in the GPA at the end of both terms:

November	3.60	April	3.60
December	4.00	May	3.70
January	3.80	June	3.87
February	3.90	July	3.85
March	4.00		

3. Executing the Training

A generic observation in the report stated that at the beginning of the year the students were very shy, had very little confidence, and were overwhelmed by the amount of material to study and the enormous job looming large before them. Therefore, "at the beginning of the year the students should be thoroughly supported politically so that their confidence would return."

The consensus was that teaching aids and charts should be simplified, well arranged and clear-cut, and more varied monthly training plans were supposed to be compiled. These measures were justified by the fact that "Our students possess very good practical skills, yet their cognitive perception is not in balance with the former." Therefore, especially at the beginning of the academic year, efforts should be made to implement diversified practical training sessions; this, however, was contradicted by the other objective requirement that those theoretical subjects that served as bases for others should be taught at the beginning of the educational process.

The leadership practices paid off in tactical studies, and the plan was to include more of this kind of training in the future. The field training sites, however, were supposed to be selected so that they would be within one or two hours' distance, because of the limited timeframe available.

Due to the short training time there were very few opportunities to conduct verbal examinations, and attempts were made to counterbalance this by inserting short written tests in the process.

One of the most important experiences regarding the exams was the finding that the principle that exams could only be taken during the exam periods was irrational and it proved to be more practical if exams immediately followed the completion of a certain larger segment of the subject matter. Yet securing preparation time for this was also considered to be necessary.

4. Training Results

The general progress of students indicated that objectives set for the first term had almost completely been satisfied, while in the second term there were significant differences among the students' achievements, especially in the area of applying the acquired knowledge.

The Staff Course mostly trained *regimental*-level or *independent battalion*-level officers, and a minority was capable of fulfilling higher and lower positions.

Both the teachers and the students made serious efforts to achieve good course results at the end of the year. 15 of the teachers achieved a GPA better than 4.0 and 11 teachers achieved a GPA better than 3.5 (with the students and subjects). In order to achieve these objectives, obviously, "The Party organization of the Academy played a great role in urging people to do a good job by constantly keeping an eye on the zeal of the teachers and the diligence of students as well. The final exam reflects the results of mutual good work led by the Party."[52]

5. The Distribution of the Course's Students' Grades per Battalion and per Branch

Branch Classes	Grades 5 and 4	Grades 4 and 3	Grades 3 and 2
I. Infantry	10	14	0
II. Infantry	11	11	1
III. Artillery	12	8	0
IV. Air Defense Artillery + Air Force + Signals Corps	8 + 4 + 4 = 16	2 + 5 + 2 = 9	0 + 1 + 0 = 1
V. Engineers + River Fleet + Armored Branch	3 + 2 + 4 = 9	1 + 0 + 2 = 3	1 + 0 + 0 = 1
Total	58	45	3
Percentage of the Total	54.7%	42.5%	2.8%

During the academic year it was troublesome that the commander could not help frequently changing the teaching assignments and the excessive—although service-related—strain on
teachers. Due to the inadequate number of teachers, a teacher had
to teach two to four subjects. It also made the work difficult that
there was not enough time to execute the commands—frequently
issued by superior authorities—thoroughly, correctly and precisely.
"Therefore, it happened that our arrangements had to be revised, or,
if they were ongoing, they had to be terminated because of a measure issued late by the Center."

6. The Disciplinary Situation

Table indicating the disciplinary situation of the academic year:

	Officer	Student	Subaltern	Rank and File
Punishment	20	47	17	161
Commendation	104	128	18	84

It was rather difficult for the students to adjust to the extremely restricted and intense new lifestyle; consequently there were
immense problems at the beginning. At the end of the academic
year exhaustion and the hot weather caused problems. Yet, all
things taken into consideration, the ratio of student punishment and
commendation was normal. It is also obvious that the teaching staff
index was much better. The punishments for teachers (20) were
allotted for tardiness, clothing deficiency, or being unshaven.

The ratio of punishments and commendations in different staff
categories was expected and realistic, although the fact that the
number of punishments in the case of the rank-and-file soldiers was
double makes us wonder.

7. The Personnel Situation

During the year there was a 15–20 percent shortage of personnel all along. The leaders of the Staff Course indicated the following justified and valid claim:

> The idea of the *Honvéd* General Staff and the Organizational Department that a large staff cannot be accorded to the relatively low number of trainees—which was the basis of rejection to our multiple submissions—was wrong. The significance of the Academy and its importance penetrating the whole *Honvédség* would have demanded that our course would be substantially endowed by personnel and other facilities definitely needed to achieve good results and to secure undisturbed learning for the students.[53]

8. Accommodation

Problems relating to accommodation only emerged when three classrooms had to be given over to the Higher Commander Training Course under Béla Király's command, starting on April 1, 1950.

The staff offices were quite crowded as well (four to six people) and it was difficult to ensure quiet teacher preparation. The student accommodation in rooms of four or five was considered to be better; however, there was a shortage of bathrooms.

The exceptional accommodation of the rank-and-file soldiers foundered on the fact that they had to give up their rooms to the offices of the developing Higher Commander Training Course. They were pushed into a makeshift arrangement in the basement of the Academy until some secondary building was built, but even after that they had to sleep on bunk beds.

Although the modern tank and gun hangars had been built, there was a lot of construction work lagging behind (a central large lecture hall, club, library, gymnasium, restaurant, kitchens, stores, and so on)...

9. Finances

Both the commander's expense account and the award account proved to be extremely small compared with the character of the institution. The fact that a large number of students were placed in lower salary categories than they had been before also became the source of serious tension. Although in time—under the pressure exerted by the Course leadership's submission—the problem was solved, it was considered that it would be more practical if this were settled in the future before joining the Course.

10. Health Care

The health of the enrolled students was good, and the signs of exhaustion—including weight loss, headaches and insomnia—cropped up only in January and February. By transforming the daily schedule, these problems decreased. The per capita unsuitability for service was eight days for students and ten days for the staff, which cannot be considered to be high.

Based on experience, the report writers proposed that a thorough and detailed medical examination should be implemented at the beginning of the new training year, and should be repeated during the year.

11. Tactical Trip

The tactical trip held between August 16 and 31, 1950, served the purpose of deepening the students' tactical knowledge and cultural background knowledge. It was recommended that issuing and practicing verbal military commands should be implemented in the future.

The workers' and soldiers' meetings proved to be very useful, although they did not reach the desired objective everywhere, due to an overcrowded program. It was recommended that the tactical trip—as long as it mobilized a large number of people—should be divided into smaller groups, and that during the trip more time should be dedicated to self-education and individual studies.

12. Summary

In spite of all the troubles and problems, the Course achieved its objectives, although this activity should be considered as experimental. They started with little experience and unsettled methods. "Yet our students' exemplary will to learn and discipline, the conscientious and diligent work of the officers, subalterns and rank-and-file soldiers assigned to the staff, the dedicated support of civilian employees, and the outstanding talent and professional knowledge of teachers, together with their unmatched care, their comradeship, unselfishness and dedication, made the Course a success. Last but not least, we could achieve our results through the organizational work and support of the Party, which was serious and thorough and always bore the objectives of the Academy in mind."[54]

48   Miklós M. Szabó

Notes

1. Honvédelmi Minisztérium, Hadtörténeti Levéltár [Ministry of Defense, Military History Archive] (hereafter cited as HM HL). 1352/eln.kfcs.1.csf.–1949 (Jan. 17).
2. László Csendes and Tibor Gellért, *Háborútól a forradalomig 1945–1955. Adalékok a magyar hadsereg történetéből* [From War to Revolution, 1945–1955. Contributions to Hungarian Army History] (Budapest, 1994), p. 94.
3. HM HL Hadiakadémia [Military Academy] (hereafter cited as HAKAD) 195/eln.–1949 (July 14).
4. HM HL HAKAD 185/eln.–1949 (July 12).
5. László Csendes and Tibor Gellért, *Kronológia a Honvédség történetéből 1945–1990* [Chronology of the History of the *Honvédség*, 1945–1990] (Budapest, 1993), p. 79.
6. HM HL HAKAD 205/Hakad.pk.–1949 (May 17).
7. HM HK HAKAD 259/eln.–1949 (July 14).
8. Lajos Móricz, "A magasabb tisztképzés magyarországi történetének elvi, szervezeti és vezetési kérdései" [Theoretical, Organizational and Leadership Questions of the Hungarian History of Higher Officer Training], *Akadémiai Közlemények* 178 (1991): 15.
9. HM HL HAKAD 080/htp.–1950 (Jan. 22).
10. HM HL HAKAD 1/24/T.Tö.tanf.–1949 (Sept. 15).
11. HM HL HAKAD 1/27/T.Tö.tanf.–1949 (Sept. 27).
12. *A Magyar Állam szervei 1944–1950 N–Z* [Bodies of the Hungarian State, 1944–1950, N–Z] (Budapest, 1985), p. 651; Csendes and Gellért, *Kronológia a Honvédség történetéből 1945–1990*, p. 89; Dr. Antal Oroszi, *A Zrínyi Miklós Katonai Akadémia történetének össze-foglalása (1947–1996). A magyar katonai vezető- és tisztképzés története (Tanulmánygyűjtemény)* [Summary of the History of the Miklós Zrínyi Military Academy, 1947–1996. The History of the Hungarian Military Leadership and Officer Training (Collection of Studies)] (Budapest, 1996), p. 251.
13. HM HL HAKAD 902/Eln.tpk. + Szü.–1949 (Oct. 12).
14. Oroszi, *A Zrínyi Miklós Katonai Akadémia történetének össze-foglalása (1947–1996)*, p. 65; Csendes and Gellért, *Kronológia a Honvédség történetéből 1945–1990*, p. 89.

15. HM HL HAKAD 1172/Eln./ HONV.10.PF.8985–1949 (Nov. 3).
16. HM HL HAKAD 1495/Eln.tpk.–1949 (Dec. 19) and 1524/Eln.tpk.–1949 (Dec. 22).
17. HM HL HAKAD 035/PK.–1950 (Jan. 9).
18. HM HL HAKAD 93/SzT.Pk.–1949. 1950 (Jan. 5).
19. HM HL HAKAD 03/T.Szü.–1950 (Jan. 9).
20. HM HL HAKAD 0166/Tö.tanf.pk.–1950 (Feb. 10).
21. HM HL HAKAD 0193/K.–1950 (Feb. 17).
22. HM HL Honvéd Gyalogság Parancsnoka [Commander of the Honvéd Infantry] (hereafter cited as HGyPk) 0977/Pk.–1950 (March 1): the research of Dr. Antal Oroszi, colonel (ret.).
23. HM HL HGyPk 0977/Pk.–1950 (March 1).
24. Csendes and Gellért, Kronológia a Honvédség történetéből 1945–1990, p. 90.
25. HM HL HGyPk 0977/Pk–1950 (March 1).
26. HM HL HGyPk 0977/Pk–1950 (March 1).
27. Csendes and Gellért, Kronológia a Honvédség történetéből 1945–1990, p. 96; A Magyar Állam szervei 1944–1950 N–Z, p. 651.
28. HM HL HAKAD 466/K.–1950 (March 21).
29. HM HL HAKAD 504/T.–1950 (March 25).
30. HM HL HAKAD 540/T.–1950 (March 28).
31. Ibid., 0386/Tö.tanf.P.–1950 (April 12).
32. Ibid., 0499/T.–1950 (April 26).
33. HM HL HAKAD 0487/Tö.tanf.P.–1950 (April 26).
34. HM HL HAKAD 0544/T –1950 (May 4).
35. HM HL HAKAD 0530/H.–1950 (May 2).
36. HM HL HAKAD 0716/P.–1950 (June 3).
37. HM HL HAKAD 0925/P.–1950 (July 4).
38. HM HL HAKAD 0925/P.–1950 (July 4).
39. HM HL HAKAD 0925/P.–1950 (July 4).
40. HM HL HAKAD 0905/Tö.tanf.P.–1950 (July 21).
41. HM HL HAKAD 0905/Tö.tanf.P.–1950 (July 21).
42. HM HL HAKAD 0905/Tö.tanf.P.–1950 (July 21).
43. HM HL HAKAD 01155/P.–1950 (Aug. 2).
44. HM HL HAKAD 01378/P.–1950 (Sept. 2).
45. HM HL HAKAD 01197/Tö.tanf.K.–1950 (Aug. 10).
46. HM HL HAKAD 086/T.–1950 (Oct. 3).
47. HM HL HAKAD 01422/Tö.tanf.P. 1950 (Oct. 4).

48. HM HL HAKAD 01422/Tö.tanf.P. 1950 (Oct. 4) d:276.
49. HM HL HAKAD 01422/Tö.tanf.P. 1950 (Oct. 4) d:276.
50. HM HL HAKAD 01422/Tö.tanf.P. 1950 (Oct. 4) d:276/a.
51. HM HL HAKAD 01422/Tö.tanf.P. 1950 (Oct. 4) d:279.
52. HM HL HAKAD 01422/Tö.tanf.P. 1950 (Oct. 4) d:282.
53. HM HL HAKAD 01422/Tö.tanf.P. 1950 (Oct. 4) d:284/a.
54. HM HL HAKAD 01422/Tö.tanf.P. 1950 (Oct. 4) d:288.

Chapter IV
The History of the Higher Commander Course
(April 1, 1950–September 18, 1950)

In the earlier chapters we became acquainted with the *Honvéd* Military Academy and with the *Honvéd* Staff Course established on the foundation of the *Honvéd* Academy on July 22, 1949. In the following year, the organization of the "Higher Commander Course" was launched.[1] Colonel General Mihály Farkas, minister of defense, took measures to set up the Course on April 8, 1950.[2]

The ministerial decree included the stipulation that "the designated commander of the Higher Commander Course is responsible for the establishment of the [institution] so that the education of the students will begin on April 17." Therefore, Commander of the Course Major General Béla Király was not in an enviable situation. Király had been transferred from his position as commander of the Hungarian Infantry,[3] which had been terminated on April 7, 1950.[4] The "request" itself was unorthodox, as General Király remembered. After the April 4, 1950, military parade, "Farkas informed me that the Staff of the commander of the infantry was going to be broken up and its sphere of action was to be taken over by the minister himself... 'We are establishing the Military Academy. You will be its commander. This will be the key task of the army from now on. You can select any officer you want for the teaching staff irrespective of their position or rank. Whatever material or financial wish you have, I will satisfy it. Turn directly to me with whatever you need so that we would avoid any bureaucratic complications.'"[5] Farkas partially lived up to his promise; his order[6] said that "The number of people displayed on the staff chart is made permanent. The appointed teachers and students need to be transferred... If the transferred teachers received higher salaries in their former appointment than what is shown in the attached staff chart, they should continue to receive the former (higher) salary."

The Higher Commander Course functioned between April and September 1950. Its training mission was the following:

As a result of the educational process the students are to acquire the bases of Marxist-Leninist theory and should be informed about the daily (current) events. They should be knowledgeable about the basic fighting principles of the modern mixed branches and they should be able to make decisions, as well as to organize mutual cooperation between the branches in war. They are to be familiar with the work of their division, army corps, and unit according to different fighting circumstances. They are to acquire the theoretical basis of, the approach to, and rules of shooting. They should be familiar with infantry weapons and they should be able to direct the application of infantry weapons in war. They are to acquire the methodology of troop training and education and the ability to engage in the military and political training of ancillary units. They are to correctly implement the most important measures of rules and regulations.[7]

The Course prepared officers for appointments as army corps commander, army corps political officer, division commander and division political officer, as well as army corps and divisional chief of staff appointments.

According to the report of May 1, from the effective number, there were still unfilled positions, in spite of the ministerial promise.

The work went ahead at full steam. Major General Béla Király, commander of the Course, and Colonel József Szabó, its political officer, submitted a draft[8] to be endorsed by the chief of General Staff, entitled "Directions to Execute the Evaluation of Students within the Higher Commander Course and to Provide Internal

Service," which was the adapted version of the formerly issued ministerial decree.[9]

The essence of the draft in terms of evaluating the students was the following:

- The evaluation of students by the teachers should be continuous.

- The exam results should be registered in the students' examination registration booklets.

- Grades are to be given within the 1 through 5 range.

- Verbal reporting should be frequent and should be graded each time.

- Topic items should be drawn at the exam, and preparation time and note-taking should be ensured.

- During the Course, summary exams should be taken from among the following subjects: general information for the officers of the *Honvédség* (hereafter cited as *technical minimum exam*); the fundamentals of attacks and defense; the organization of the protection of an infantry division under normal and extraordinary circumstances; organizing the attack of an army corps.

- The relevant party should be informed about the result of the exam, and the result should be registered in the exam registration booklet.

- On completing the Course, the students need to take a summary (final) exam before an extended committee.

The essence of the draft regarding internal service and internal order was the following:

- The rules of the Army Regulations are determinative.

- Since the Higher Commander Course is accommodated in the barracks of the *Honvéd* Staff Course (address Hungária körút 9–11, today called the University of National Defense), the bulk of the services is offered by them.

- Regarding the sphere of authority for commendation and punishment, the following are the competencies: the commander of the Course has those of an *army corps commander*, the head of the military department has those of a *regimental commander*, and the independent training sub-department head has those of a *battalion commander*. The teachers—irrespective of their rank—are superior to all the students.

Several important documents issued sequentially emphasized the speed of the Course's organizational process. The report submitted to the chief of General Staff on May 10, 1950, fell into this category. Its subject was "Report and Request for Conceptual Decisions regarding the Establishment of the Higher Commander Course."[10]

All in all, Major General Király decided that the Course was *necessarily functional* but the recruitment should be completed by April 15.

Major General Király sought advice as to whether the academic year opening should be ceremonious, and if it were to be festive, whether it should be the minister or the chief of General Staff who participated in it. If the decision was for a ceremonious opening act, Király proposed April 16, 1950, between 10:00 a.m. and 11:00 a.m., so that teaching could begin on April 17 according to previous decisions.

As far as the clarification of the conceptual questions was concerned, the commander of the Course suggested that four training periods would be accepted:

1. The *"theoretical period"* (three weeks), with the objective of offering theoretical preparations for the concrete and application-oriented discussions (group activities). This period was to be concluded by an exam, for the execution of which he requested the release of 8 hours from the reserved 60 hours.

2. The *"period of attack"* (ten weeks), when all the topics related to regimental, divisional and army corps attacks were to be taught within the framework of concrete and application-oriented discussions, which were to be concluded by an exam.

3. The *"period of defense"* (seven weeks), in order to review the defense activities of the above units and to conclude the period with methods similar to the period on attack.

4. The *"summary period"* (two weeks), when the students were to be informed about questions in connection with field army leadership.

Major General Király expected that great attention would be paid to the familiarization of the students with the different branches and their military techniques through branch-performed technical shows and through afternoon visits to troops placed in the Greater Budapest area. He also reported that the Soviet advisors[11] had offered to place themselves at the students' disposal during the evening consultations.

The leadership of the Course sought the higher authorities' agreement to the idea of establishing two classes ("A" and "B"), whereby students coming from higher ranks and positions would be proportionately distributed between the two. The syllabus would be the same for both groups. Comprehensive topics would be presented at consolidated lectures and discussions, and debates and seminars would be held in separate classes, while the evening review sessions would take place in four study circles (six to seven people each).

The fact that the Course was really a higher-level officer training course was reflected by the fact that Colonel General Mihály Farkas visited the Higher Commander Course on May 5.[12]

The situation with the number of participants of the Higher Commander Course had slightly improved by June 1, 1950, and there were even supernumerary people in certain areas. The regular/effective number showed the following picture on June 1, 1950: 23/24 officers (of whom there were 2/2 political officers), 1/1 subaltern, 0/1 *honvéd*; 30/24 officer students and 20/25 civilian employees, 70/75 people altogether.[13]

Major General Béla Király

The commander of the Course and its political officer submitted an informative report on the effective force and training questions to the minister of defense on June 22.[14] According to the submission, the following military served as the Course staff: 1 general and 13 teachers (1 infantry colonel, who was also the deputy commander; 1 artillery colonel and 1 armored colonel; 3 infantry lieutenant colonels and 1 artillery lieutenant colonel; 3 infantry majors; 1 engineering captain; 1 signals troops first lieutenant and 1 infantry first lieutenant). Furthermore, the positions of 1 artillery teacher, 1 engineering teacher, 1 anti-aircraft artillery teacher, and 1 air force teacher were also established but remained unfulfilled.

The most important piece of information, however, was the fact that the students were of the following ranks: 2 generals, 6 political officers, 2 armored colonels, 1 infantry colonel, 3 infantry lieutenant colonels, 8 infantry majors, 3 armored majors, 2 artillery majors, 1 anti-aircraft artillery major, 1 artillery captain and 1 anti-aircraft artillery captain.

	Per the report on effective force of July 1, 1950[15] (regular/effective)
Officer	23/25 (including 2/2 political officers)
Subaltern	1/ 2
Corporal	1/ 0
Honvéd	0/ 1
Civilian Employee	20/26
Student	30/26

The training objectives of the Course were specified in the Information Report[16] as follows:

- To develop and improve the generals' and officers' knowledge and skills, in order to organize, prepare and secure the tactical aspects of modern mixed-branch formations;

- To make students acquire practice in commanding mixed-branch units and formations;

- To improve the theoretical and methodological skills of the staff in training troops;

- To familiarize the work of divisional and army corps staff in different combat processes;

- To teach the acquisition of the theoretical foundations of ballistics and to teach the application of infantry weapons in battle;

- To teach the methodology of troop training.

The Course trained participants to provide service in divisional and army corps level positions as commanders, political officers and chiefs of staff. After completing the training on September 17, 1950, the students were to occupy these positions.

In order to understand the details of the training taking place in the Higher Commander Course, three factors need to be reviewed:[17]

- The role of Soviet advisors;

- The lack of regulations and related necessary provisions as the greatest impediment of the compulsory transformation into a Soviet system;

- The introduction and advantages of the "cabinet chamber," a successful Soviet method.

The Soviet advisors overran Hungary from the fall of 1948 onwards. As was elaborated in Chapter II, Stalin would have been ready to inflict aggression and launch the war in order to solve the Stalin–Tito conflict. The task of the advisors both in Hungary and in all the satellite nations was to help organize huge armies and to have prepared them by 1950 to participate in the war against Yugoslavia.

Where trained officers were in command, the advisors gave only "advice," but everybody knew that it was life-threatening not to accept it. Where, however, the commander was a newly appointed "cadre," a reliable member of the Communist Party (Hungarian Workers' Party), and had absolutely no idea of what military leadership was, the advisor himself drew up the commands, which had to be signed by the Hungarian commander after they had been translated into Hungarian.

The members of the faculty of the Higher Commander Course were all, without exception, trained officers who had planned and completed their work themselves. The majority of the Soviet advisors who were connected to the Course were highly educated General Staff officers who were willing to help the faculty to understand the necessary changes. But as always, there were exceptions. There were advisors who always found fault and suspected sabotage or conspiracies everywhere. This could entail life-threatening circumstances for those unable to conform to them. The most dangerous was Guard Colonel Voloshin, whose behavior is worth demonstrating by quoting Commander of the Course Major General Béla Király's recollections.

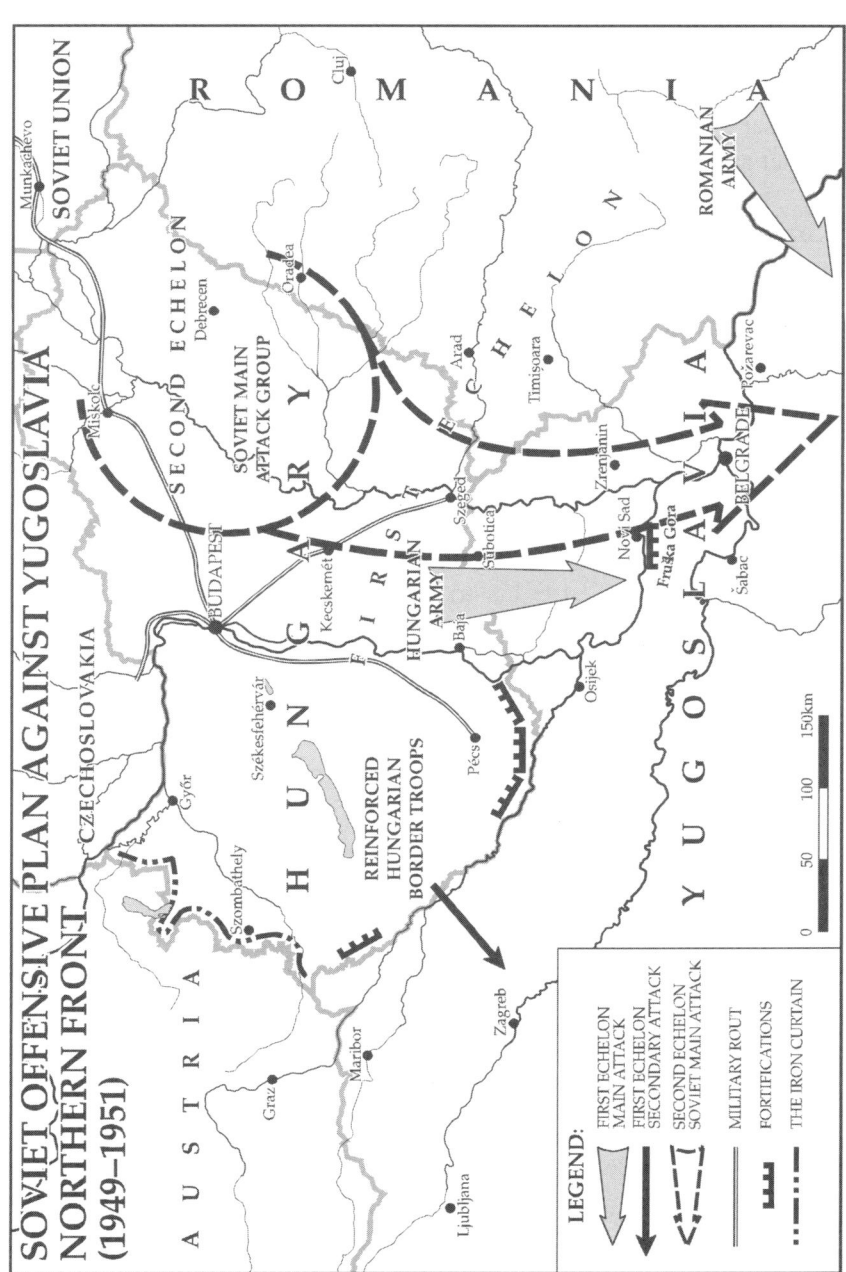

SOVIET OFFENSIVE PLAN AGAINST YUGOSLAVIA
NORTHERN FRONT
(1949–1951)

LEGEND:

FIRST ECHELON
MAIN ATTACK

FIRST ECHELON
SECONDARY ATTACK

SECOND ECHELON
SOVIET MAIN ATTACK

MILITARY ROUT

FORTIFICATIONS

THE IRON CURTAIN

By January 1951 most of the advisors had already worked at the Course; their commander, Samsonov, had served as a cadet in the Tsar's army and had received the rank of major general in the Soviet army.

At the end of December 1950 General Samsonov told Béla Király that on the following day he would introduce him his newly arrived tactical advisor Guard Colonel Voloshin, Hero of the Soviet Union.

Like all Soviet advisors, he showed up for the introduction in a brand-new uniform, but it soon became apparent that he cared not a whit about his appearance: the layers of dandruff on his shoulders appeared to get thicker by the day. Occasionally, after a field exercise, the same mud was still stuck to his boots for several days. He was sickening just to look at.

Barely a week after his arrival my aide-de-camp made a discovery that appeared unbelievable: namely, that Colonel Voloshin, rather than the major general assigned to us from the beginning, was the leader of the advisors. When I expressed my doubts, he said, "We have noticed that each morning it is not Colonel Voloshin who goes to report to the office of the major general on the first floor, but the other way around: it is the old general who treads his way upstairs to the second-floor office of the colonel with a bunch of documents in his hand. We even overheard a loud exchange of words, and the louder voice did not belong to the general."

My aide-de-camp had no vested interest in misleading me; therefore I acknowledged his report that Voloshin had come with some kind of special mission from an unknown branch of the Stalinist regime. I was sure about one thing: I had better be careful. Although his position was that of an advisor in tactics, he had his nose in everything. One day we were discussing world affairs during a lecture. The topic was introduced by a guest lecturer, to be followed by a "debate." Voloshin took part in the debate, and lambasted the imperialists.

"They slander the Soviet institution of political commissars as an instrument of oppression; yet I spent half a year towards the end of the war as political commissar at the front, and I only shot three cowardly Soviet soldiers, all three of them for good reason."

It became increasingly clear that, no matter what Soviet agency Voloshin represented, his assignment was not to advise in matters of instruction or tactics, but rather to make sure that instructors were true to the Soviets. Loyalty not to communism, but to the Soviet Union, was the most important thing [for Voloshin]...

In June 1951, beside himself with rage, Voloshin brought a poster with him from the wall of the so-called "cabinet room."... The transparency illustrated an offensive by a reinforced regiment, showing the battalions arranged in two echelons, the supporting artillery and the tanks, as well as the observation post of the commander.

"Here is a prime example of anti-Soviet propaganda! We must take measures against the culprit!" Voloshin said in a commanding voice.

"I don't see anything anti-Soviet on this sketch," I answered.

"Which way is this regiment attacking?"

"To the right."

"Yes, but if you look at the continents on the world map, what is to the right? The east, isn't it? And what lies to the east of Hungary? The Soviet Union. Therefore, this sketch illustrates an attack against the Soviet Union. It is anti-Soviet propaganda. Take immediate action!"

The three Soviet soldiers shot by Voloshin came to my mind. Perhaps the reason for their punishment was no worse than what the professor who designed this sketch had done. It would have made my position secure if, in the spirit of Soviet justice, I had had the individual arrested regardless of his innocence. Of course, I was not inclined to do that, so I said, "What you are holding in your hand is not a map, but a sketch. The cardinal points are not indicated. If you turn the sketch around, the regiment will be attacking to the left, that is, towards the west. I don't feel that any measures are necessary."

Without another word, he slammed the sketch to the ground and left.

A few weeks later he objected because our course in military geography, that is, the studying potential theaters of operation, included the Carpatho-Ukraine.

"That region is an integral part of the Soviet Union. If you teach it as a potential theater of military operations, then you are assuming that the imperialists might be able to tread holy Soviet soil with their filthy boots. We must put a stop to this at once! Hungarian officers have to focus on Yugoslavia as a theater of operations. If they are awakened in the middle of the night and asked how deep the River Sava at Šabac (Serbia–Montenegro) is, they must be able to answer. Obviously, this is anti-Soviet propaganda. Have the culprit arrested!" demanded Voloshin.

I was in the same situation as earlier. I would have to send an innocent man to the gallows. Moreover, it was my lifelong friend, Sándor Zsilinszky, who was the "culprit." I did not have to think about the answer.

"I shall not arrest anyone! I refuse to prepare a Hungarian officer of the General Staff for just a single theater of operations. I am training them for their whole life."

Two days later my secretary, who lived only a few doors down from Zsilinszky, told me in tears that, in the morning, on her way to the office, she had seen the lieutenant colonel leaving the tobacco shop, between two civilians pushing him into a vehicle parked nearby. I immediately sent my aide-de-camp to Zsilinszky's place. He came back saying that the lieutenant colonel had left in the morning and had not returned since. Nor did he show up the next day; so the security servicemen attached to the Academy sent out a circular about his being absent without leave, of which they sent me a copy, but all of us knew where the matter stood. Soon I was arrested too.[18]

As was mentioned before, most of the Soviet advisors behaved diametrically differently from Voloshin. They did not conceive of the Hungarian officers as the enemy but wanted to help them. In this regard, Lieutenant Boiko, Hero of the Soviet Union, showed an outstanding human disposition. He was head advisor to the minister of defense. The fact that Lieutenant General Boiko visited the Course every week also demonstrated that the Course was the first priority of the army. He certainly offered well-meaning pieces of advice, and if he did not agree with something, he expressed it in a civil manner and made suggestions as to what the right solution would be. On each occasion his parting words were that the Course's commander was doing everything in his power, and Boiko agreed to continue the work without any fundamental change.

This happened in connection with the lack of Soviet regulations and related activities. Based on the above, it must be evident that the Hungarian army needed to be transformed and trained according to the Soviet regulations. The translation of the necessary regulations had already started; however, final Hungarian copies were not available yet. Therefore, the commander of the Higher Commander Course made recourse to a very dangerous method: he used the completed but not finalized regulations and made copybook-sized publications, in other words bulletins, in order to facilitate the teaching of the most important subjects. Lieutenant General Boiko took the drafts of the most important topics with himself and shared his opinion about them a few days later. Quite

contrary to Voloshin's attitude, if Boiko found something of which
he disapproved, he recommended modification and not hanging.
All in all, the bulletin program proved to be very useful in the work
process of the teachers. Boiko spoke appreciatively about the
Course both to Mihály Farkas and to the first person of the Party
and the government, Mátyás Rákosi. General Király remembered
this as follows:

> ... He must have reported his positive impression to Rákosi as
> well, I felt, because some time during the fall of 1950 I was
> summoned to appear in front of Rákosi, at Party headquarters.
>
> This kind of invitation always worried those invited,
> because one never knew if the trip might not end up in the jails
> of the Security Service. But this time I was led straight to the
> office of Rákosi, without the usual ceremonies. I was received
> in the spirit of a royal audience from the time of the Imperial
> and Royal Army; although "his majesty" shook my hand, he
> did not ask me to be seated, and let me leave after a few words.
> I had time enough to observe the furnishings in his office. Next
> to his desk stood two American-type metal filing cabinets,
> with Stalin's bust on top of one of them. On the right, at the
> end of a long conference table, I saw a heap of our bulletins
> prepared for the [course]... Pointing at them, Rákosi said, "I
> cannot say that I have read every word that you all have writ-
> ten, but enough to know that you are doing good work. This is
> the way to build the People's Army!"
>
> That was the end of the audience. I was able to continue to
> work for a short while under the impression that there was no
> immediate danger.[19]

Issuing bulletin-format professional articles to facilitate the
teachers' work was the idea of the leadership of the Higher
Commander Course. Another significant scientific aid to help students to understand the materials of the lectures was the invention

of "cabinet chambers." They were installed by the commander of the Course at the suggestion of the Soviet advisors. Both methods were realized during the life of the Higher Commander Course and continued to function at the *Honvéd* Academy founded during the fall of 1950.

The essence of the "cabinet chamber method" was that on the walls of these rooms there were large, poster-like sketches. The themes of the sketches were in harmony with what the students heard from their teachers during the lectures, and they could study them after lectures in the cabinet rooms.

The commander of the Course and of the *Honvéd* Academy, Major General Béla Király, described the practical implementation of these methods as follows:

The sketches were changed weekly according to the actual material taught. The drafts were precise and artistically appealing. I appointed a drawing brigade from the students of the College of Fine Arts. The fine arts students implemented the drafts drawn up by professional cartographers under the direction of professors of tactical studies into beautiful artistic forms. I was able to pay these fine arts students from my unlimited budget and the pay seemed fabulous to them, which they returned with enthusiastic contributions. I visited this group of artists on a daily basis. We became friends and enjoyed the beauty of their creations together.

The Tactical Department taught a different topic every week. The topic was "attack" during the week I first collided with Voloshin.[20]

To understand the composition and work of the Higher Commander Course and the *Honvéd* Academy and the development of the People's Army, two issues are worth pointing out in the following pages:

- The relationship between students and teachers;
- The typical study results and careers of some students who received important positions in the People's Army after completing their studies.

The fact that some teachers were of lower rank than the students—who were generals and high-ranking officers—caused difficulty in the students' and teachers' relationship. Several teachers did not appropriately criticize such students if they made mistakes. But it is evident from the "summarized grading and qualifying register" that the evaluation of students was appropriate all in all.

The following passages are excerpts from these documents.

In the case of Major General Mihály Szalvay,[21] his unending diligence and practicality were acknowledged but his weaknesses were also pointed out. He received the following academic evaluation based on his grades—six 4s and one 5—received for subjects regarding "The basic tactical principles of mixed branches, tactical securing of troops, and tactical and technical basic data of mixed branches": "Be careful in your further studies… emphasize memorizing military expressions, because military organization should be as logical as the ideological one. You have studied extremely diligently."[22] Grades 4 and 5 at regimental or divisional exams were also qualified as communicating the following message: "Your discipline and diligence are exemplary—you can set an example. You are very practical. You are extremely logical in practical situations. You should emphasize staff service." The qualification conferred in the exam discussions especially proved Lieutenant Colonel Sándor Zsilinszky's high expectations, which were based on principles and were very definitive, although these were already hard times for a former Hungarian Royal *Honvéd* officer. On the whole, his ideological preparedness, acquired during his several-decade-long involvement in working-class movement activities, his pragmatism, acquired as a battalion commander in the Spanish Civil War, and his sometimes irregular solutions were acknowledged, but his weaknesses and significant shortcomings in the areas of the state-of-the-art military science and commandership activities were also pointed out.

There were similar experiences with another "worker cadre," Colonel Lajos Gyurkó,[23] who had the following statements in his academic record: "It is of the utmost importance that you do not get lost in minute details and that you search for coherence. Always search for the leadership part and search for the essential element in the depth of every question."[24] As far as the topical area of "the attack and defense of an infantry regiment" was concerned, "You understand the essential element, you pay attention and you are active. You successfully struggle against getting lost in petty details." The summary of his exam in connection with the infantry division was that "Your drawing skill has improved well; continue to work on it. This is the virtue of a commander if he can lead and plan in drawings... In terms of organization, great strides have been taken. Your work map is very good... You do not calculate the time factor sufficiently. Regarding balancing training you are a radical, yet you often slide over to areas where side-directions are brought to the foreground more than is needed. These phenomena demonstrate that you do not fully consider every decisive problem... You make very good critical utterances; continue to grow in this direction."

It was difficult for him, however, to cope with the demands of the exam discussions. Its evaluation was the following:

"I. Decision... Result: 3
II. Command... Result: 3."

Partly similar and partly different problems emerged in connection with Major General Gyula Uszta,[25] whose "student behavior" sometimes reflected the self-confidence of a former partisan commander and that of a responsible state and Party functionary. In his final grade and academic summary,[26] containing grades 4 and 5, the following evaluations can be found: "His unlimited diligence is commendable. He is also very active. *Yet we need to warn you to always wait until the teacher finishes his sentences, and you should not disturb him with your interruptions.* You should be more systematic; concentrate on the essential element, and do not get lost in details." In connection with exams on tactics there were the following statements: "Very active. Organizes study groups with his class very well... He has set out on the road to look for and eventually find the decisive point of a situation. *Yet it could also be ascertained that, when confronting difficulties, he did not set out to solve the problems but criticized the causes. This is not acceptable. Correct your mistake in the future.*"

In the divisional exam, the board of relevant parties already established that "His drawing skills have substantially developed and he has made further advancement in visualizing essential problems." The evaluation of the exam discussion, the crown jewel of the whole evaluating process, was the following:

"I. Decision... Result: 4

II. Command... Result: 3.5."

The essence of the epoch is well manifested by these evaluations: the evolving new society did not dare to trust the "class enemy" old officers, who were well trained and had immense experience. Yet the politically reliable "new cadres" mostly came from less educated and cultured social strata, for whom learning took up a great amount of intellectual energy, was strenuous, and often became a health hazard.

Perhaps that was the reason why Colonel Pál Maléter[27] could not be as self-assured as Major General Uszta. Maléter had also returned as a partisan from World War II, but during the months of extensive show trials his background with training at the Ludovika Military Academy must have weighed heavily on him. His evaluation was the following: "In his profession, he should search for the essential element, as he has done before. This is his strong suit in this field." He was among the ones with an "outstanding" evaluation in tactical studies, since he completed eight of the nine tactical exams at grade 5 and only one at grade 4. In regimental tactical studies he received the following warning from his examiners: "You should develop your essay-writing skill, be more precise in this field, and try to systemize much better. You are very good at collective work." During the course of the instruction on the "infantry division" subject matter, Colonel Maléter must have committed a major "crime," because the following was stated in his academic register: "There were situations when he did not pay attention... Yet it can be established that he underwent a major development." Later, during the courses related to division and army corps subjects, his evaluation was radically different. "His capacity to obtain a comprehensive view is very good, and he has shown serious development in this field. He is very active. His drawing skills have improved very nicely. He is striving to reach the essential element of the matter. His participation in the activities is mindful and active. He should pay more attention to his written work. He is advised to summarize his thoughts in a few lines, so that his presentation would be systematic. He is very active and observant, and he does not miss the smallest mistake." The exam discussion was successful:

"I. Decision... Result: 4
II. Command... Result: 4."

The talented young people also appeared and showed their strengths in establishing their careers. For example, First Lieutenant Imre Kerekes,[28] who, out of 18 exams, passed 14 exams at 5 and 4 exams at 4. Accordingly, his academic evaluation was enviable: "He is extremely thorough, gets to the core of things, is diligent, and has a military presence. He has made giant steps forward with regard to acquiring academic knowledge. He should actively operate in this direction."[29] At the beginning he was handicapped by his youth and by lacking experience. Therefore, in the first exam—regimental subject matter—his evaluation said the following: "He is not practical enough... He does not show enough interest; he should be more active." The warning worked, because in the divisional fighting activity he mostly had 5s—as opposed to the earlier 4s—and his academic evaluation also changed. It was stated that "It is commendable that in spite of his injury he actively participates in the activities... He has serious, definite and thorough knowledge. His diligence should not let up, and he should not become conceited..." He was able to intensify studying for the army corps exam, since "His knowledge has further improved. However, his participation in the collective work should be more intense. His activity has increased, but he should increase it more. His commanding technique is below his general knowledge and he should improve it." All in all, his intellectual capacity and diligence could make up for the initial inexperience, since he belonged to the precious few who left the exam discussions with an "excellent" qualification:

"I. Decision... Result: 5
II. Command... Result: 5."

Major Lajos Tóth[30] was also an outstanding member of the group. He passed 16 of his 19 exams at 5, two at 4 and one at 3. His academic evaluation was also very positive, and he was advised to observe the "principle of gradation": "He should apply his theoretical preparedness to the practical fields. He should be more collectivism-centered. He should deepen the outstanding results of the technical minimum, and this should be connected to practice." Concerning divisional exams, "He is very active, and sees the core of things clearly and correctly. He works quickly and his drawing technique is very good. He is an original thinker and is precise. Sometimes one has an impression as if he were not present in class." And as far as the divisional studies subject matter was concerned, the following was said: "Very diligent, he develops in the right direction, and budgets his time well. He should not deal with details. He should try to grasp the essential element."[31] Unfortunately, however, the same happened to him as happened to several of his excellent fellow students: he panicked under the great pressure of the "great opportunity," and therefore the exam discussion was unbecomingly bad:

"I. Decision… Result: 3

II. Command… Result: 3.5."

The students' "ordeal," however, did not end with the exams at the end of July. Studying hard continued, followed by harsh examinations. Major General Béla Király, commander of the Course, and Colonel József Szabó, its political officer, submitted a report to the relevant authorities on the preliminary evaluation of the graduating students.[32] They indicated that there were three more exams to come, and therefore they would only be able to take a final position afterwards. This, however, did not prevent them from making the following suggestions from a *general* viewpoint:

- Students who graduate from the Higher Commander Course should possibly be placed in the following positions: division commander, political officer, or army corps chief of staff. In this way it would be possible to pursue a planned cadre development, since those whom the new graduates replaced in their positions could be enrolled in this course.

- It was only in exceptional cases that the appointment of students graduating from the Higher Commander Course to divisional chief of staff positions could be recommended, because the ones graduating from the Staff Course were more prepared for that. Its explanation lay in the fact that the thrust of the training in the Higher Commander Course was providing commandership functions and not staff work.

Király and Szabó made concrete suggestions about students—who were also mentioned above and who were also placed in important positions later—as follows:

Major General Mihály Szalvay: infantry division commander;
Major General Gyula Uszta: army corps commander;
Colonel Pál Maléter: army corps chief of staff;
Colonel Lajos Gyurkó: division commander;
Major Lajos Tóth: armored army corps chief of staff or
  Academy staff commander or head of the Academic Affairs
  Department;
First Lieutenant Imre Kerekes: army corps Political
  Department head.

There was a different kind of suggestion as well. Captain
László Baranyai did not have any military preparation at all, and
joined the Course a month late. He was not suitable for a higher
commandership position, but since he was talented, it was suggested that he should be sent to the Academy, starting during the following fall.

By the end of September all the ordeals were over, the exams
had been completed with relatively good results, and the Course
had come to an end. The "Summary Academic Results"[33] testified
to the successful educational and training work, since 19 achieved
a 5.0, 9 achieved a 4.0, and only 2 ended up with a 3.0. However,
this picture is not quite accurate, since in several instances the 4.06
or 4.13 was rounded up to 5.0. Yet it can be stated that the Course—
including all its virtues and faults—had fulfilled its mission: the
higher-level commandership and staff service training had begun in
post-World War II Hungary.

There was not much time left for complaining, since a submission[34] indicating that the Higher Commander Course would be
terminated by September 18, 1950, was complete by September 1,
1950. The submission also included the stipulation that the *Honvéd*
Academy, to be set up on September 25, would take over the tasks
of both the Higher Commander Course and the Staff Course.

Notes

1. "Parancsnokság és az akadémiavezető szerveinek története" [The History of the Different Bodies of the Commandership and the Commanders of the Academy], *Akadémiai Közlemények* 120 (1982) 1: 13; Dr. Lajos Móricz, "Adalékok a Zrínyi Miklós Katonai Akadémia történetéhez" [Additional Material on the History of the Miklós Zrínyi Military Academy], *Honvédelem* 11 (1987): 80; Dr. Antal Oroszi, *A Zrínyi Miklós Katonai Akadémia történetének összefoglalása (1947–1996). A magyar katonai vezető- és tisztképzés története (Tanulmánygyűjtemény)* [Summary of the History of the Miklós Zrínyi Military Academy, 1947–1996. The History of the Hungarian Military Leadership and Officer Training (Collection of Studies)] (Budapest, 1996), p. 251; Dr. Antal Oroszi, "Ötven évvel ezelőtt..." [Fifty Years Ago...], *Új Honvédségi Szemle* 12 (1997): 65; Dr. Gyula Balogh, *A hadiakadémiai képzéstől a szervezett doktori (PhD) képzésig. Tanulmány* [From the Military Academy Training to the Organized Doctoral (PhD) Training. A Study] (Budapest, 2001), p. 9.
2. Honvédelmi Minisztérium, Hadtörténelmi Levéltár, Hadiakadémia [Ministry of Defense, Military History Archive, Military Academy] (hereafter cited as HM HL HAKAD) 0463/Mag.pk.tanf.–1950 (April 24).
3. László Csendes and Tibor Gellért, *Kronológia a Honvédség történetéből 1945–1990* [Chronology of the History of the *Honvédség*, 1945–1990] (Budapest, 1993), p. 97.
4. See the Biographies of Key Personalities.
5. Béla K. Király, *Honvédségből Néphadsereg: személyes visszaemlékezések, 1944–1956* [From *Honvéd* Army to People's Army: Personal Reminiscences, 1944–1956] (Paris – New Brunswick, NJ, 1986), p. 177.
6. HM HL HAKAD 0463/Magasabb Parancsnoki Tanfolyam [Higher Commander Course]–1950 (April 24).
7. HM HL HAKAD–1950/T.539. d: 210.
8. *Ibid.*–1950 (May 4).
9. HM HL HAKAD Törzstiszti Tanfolyam [Staff Officer Course]–1950 (April 26).

10. HM HL HAKAD 0584/Magasabb Parancsnoki Tanfolyam–1950 (May 10).

11. In order to facilitate the efficient takeover of the experiences and practices of the Soviet military science education and officer training, Soviet military advisors, led by Major General Samsonov, arrived at the school in early 1950. For details, see Dr. Lajos Móricz, "A magasabb tisztképzés magyarországi történetének elvi, szervezeti és vezetési kérdései" [Theoretical, Organizational and Leadership Questions of the Hungarian History of Higher Officer Training], *Akadémiai Közlemények* 178 (1991): 17.

12. HM HL HAKAD 0716/P.–1950 (June 3).

13. HM HL HAKAD 0217/2.–1950 (June 1).

14. HM HL HAKAD 0860/Magasabb Parancsnoki Tanfolyam–1950 (June 22).

15. HM HL HAKAD 0217/3.–1950 (July 1).

16. HM HL HAKAD 0860/Magasabb Parancsnoki Tanfolyam–1950 (June 22).

17. Béla K. Király, *Wars, Revolutions and Regime Changes in Hungary, 1912–2004. Reminiscences of an Eyewitness* (New York, 2005), pp. 215 ff.

18. *Ibid.*, pp. 245 ff.

19. *Ibid.*, pp. 244–245.

20. Király, *Honvédségből Néphadsereg*, p. 181.

21. See the Biographies of Key Personalities.

22. HM HL HAKAD 01111/1. Magasabb Parancsnoki Tanfolyam–1950 (July 27).

23. See the Biographies of Key Personalities.

24. HM HL HAKAD 01111/1. Magasabb Parancsnoki Tanfolyam–1950 (July 27).

25. See the Biographies of Key Personalities.

26. HM HL HAKAD 01111/1. Magasabb Parancsnoki Tanfolyam–1950 (July 27).

27. See the Biographies of Key Personalities.

28. After fulfilling different higher positions, Imre Kerekes became the first secretary of the Hungarian Socialist Workers' Party of the Miklós Zrínyi Military Academy in the 1970s, and was one of the most decisive personalities in modernizing education.

29. HM HL HAKAD 01111/1. Magasabb Parancsnoki Tanfolyam–1950 (July 27).

30. See the Biographies of Key Personalities.
31. HM HL HAKAD 01111/1. Magasabb Parancsnoki Tanfolyam–1950 (July 27).
32. HM HL HAKAD 01339/1. Magasabb Parancsnoki Tanfolyam–1950 (Aug. 29).
33. HM HL HAKAD 01111/1. Magasabb Parancsnoki Tanfolyam–1950 (July 27).
34. HM HL HAKAD 01349. Magasabb Parancsnoki Tanfolyam–1950 (Sept. 1).

Chapter V
The National Defense Academy's Foundation and First Academic Year
(October 10, 1950–September 30, 1951)

The National Defense Academy was founded just three months after the Korean War broke out on June 25, 1950. Its establishment was a result of the Stalin–Tito conflict, as pointed out in Chapter II. If Mao Zedong had proven to be right in saying that the West (in particular the United States) was a "paper tiger" that would not intervene in violent Communist expansion of the type attempted in Korea, the military might of the Soviet camp—including Hungary—would have been turned against Yugoslavia. This made military potential a key issue in the countries of the Soviet bloc and gave all the requirements of the National Defense Academy immediate priority.

Furthermore, there was not just the Stalin–Tito conflict to consider. This was the period when the Cold War heightened and the world was divided into two camps, some of the antecedents being the initiation of the Marshall Plan (on April 3, 1948), the foundation of the Council of Mutual Economic Assistance (Comecon, January 25, 1949) as a key institution of the Soviet bloc or "socialist world system," the foundation of the North Atlantic Treaty Organization (NATO, April 4, 1949), the explosion of the first Soviet atom bomb (September 25, 1949), the signature of the Sino-Soviet treaty of friendship, alliance and mutual assistance (February 14, 1950), and the passage of the McCarran Act in the United States, ordering the registration of "communist organizations" (September 23, 1950).

The spread of the Cold War is quite apparent in the frequent and increasingly unrealistic ideas for developing the Hungarian armed forces. The 1949 target of 30,000 men was still within the country's capabilities, but the same could hardly be said of the "Bem" plan for the organization of the Hungarian armed forces calling for 45,000 men, let alone the strength of 144,896 men at the end of 1950, which also broke the limits set in the 1947 Paris Peace Treaty.[1]

Although Hungarian officers' training institutions (including the Army Staff Course and the Higher Commander Course) had turned out 11,000 officers by October 1950, just as many old officers had been dismissed.[2] Even so, there were 13,361 officers serving at the end of the year. The order of organization dated September 1950 spoke of an army of 120,000, including 15,000 officers, and the need to improve quality was also appreciated, although it was not a prime consideration in the tense international situation of the time. The 11 dismissed generals were replaced in a numerical sense by new appointments, but 8 of the 24 generals, including Colonel General Mihály Farkas, the minister of defense, had not attended a military academy, 7 had not completed a secondary education, and 3 had only completed the eight grades of elementary school. The position among the other officers was no better, for the vast majority had only completed elementary school and a third of them had only 1–6 years of education behind them.[3]

This was the historical background against which the National Defense Academy was founded and obliged to operate.

By mid-1950, it was being widely realized that "course-type training" of various kinds "cannot stand in for Academy education."[4] So Major General Béla Király, commander of the National Defense Academy but still in charge of the Higher Commander Course at the time, and Colonel József Szabó, its political officer, responded to an order from the chief of General Staff, and on September 1, 1950, submitted, in Hungarian *and Russian*, a plan of action and general training draft on establishing the Academy to the minister of defense.[5] It was thought necessary to point out that this document also embodied the view of the operations chief and his advisor, Major General Dmitri Leonidovich Kazarinov. They requested guidance on whether they should include English and French in the curriculum as compulsory subjects, alongside the compulsory Russian language, recommending that "the languages of the imperialist powers should not be taught as compulsory subjects in the following academic year."

The plan of action also recommended that the Operations (I) Group Section of the General Staff should issue, on the following day (September 2, 1950), an order establishing the National Defense Academy (HAKAD). This should stipulate that the Higher Commander Course conclude on September 18 and the Staff Course on September 25, and that the Academy be designated as the successor to both. Establishment of the new training institution should be commenced immediately by the commander of the Academy, who would be empowered to make use of the personnel and equipment of both courses.

In the training tasks and guidelines, three sections were envisaged:

- A two-year Academy Faculty (with four classes, for infantry, armored, artillery and air force units);

- A one-year Academy Faculty (one class with uniform training, for mixed branches);

- A Regimental Commander Course (seven classes for various types of branches).

A detailed training draft was promised for October 1, and work began immediately to gather materials for lectures and group drills. The guiding principle was that the backbone of the staff would come from the teaching staff of the two courses being incorporated into the Academy, but these would have to be augmented. At least three students of the Higher Commander Course would be required in chief of staff and head of faculty posts, and each faculty would need as a teacher at least one Staff Course student or officer returning from the Soviet Union. It was recommended that the teaching staff be assembled by September 20.

The basic principle in choosing the student body was that competitive applications would be advertised for the two-year section and that the Chief Section of Personnel of the Ministry of Defense would assemble—with a *50–100 percent safety margin*—the contingent from which the Academy Command would choose the student body by *examination*. For the one-year section, the head of the Chief Section of Personnel had to assign the students from those in posts in army corps and army corps staffs, division and division staffs, and similar posts in high commands. Students on the Regimental Commander Course would likewise be assigned by the head of the Chief Section of Personnel based on recommendations from the Branch Commands and the leaders of sections and the Chief Section in the Ministry.

The measure setting up the Academy was signed by Colonel General Mihály Farkas, minister of defense, on September 14.[6] "I order the establishment of the National Defense Academy (HAKAD) for the conduct of higher military training," it stated. "The function of the HAKAD is to teach, on the foundations of Marxist-Leninist theory, the military science of Stalin, victor of the Great Patriotic War and leader in military sciences." The Academy was placed under the chief of General Staff. With some modifications, the recommended structure of three faculties was accepted "permanently":

Branch I: HAKAD two-year section;
Branch II: One-year Higher Commander Course;
Branch III: One-year Regimental Commander Course.

Branch I faced the task of so training and educating its students as to turn them by the autumn of 1952 into officers knowledgeable in all branches and capable of serving as regimental staff and chief of regimental staff, and of performing similar functions in central administrations. They were divided into four groups:

infantry, armored, artillery and air force. It was recommended that the emphasis in the training should be on general tactics, but the students also had to have good specialist knowledge in their own branch. All four groups had to be taught the principles of commanding a division, receive guidance in the command of an army corps, and gain a general knowledge of the operations of an army.

All groups in Branch II, irrespective of their military branch, were to receive the same training and education, based on the training material in the still operative Higher Commander Course. This covered the duties of the following:
- the commander of an army corps;
- the political officer of an army corps;
- the chief of staff of an army corps;
- the commander of a division;
- the chief of divisional staff.

Students in Branch III had to be trained and educated in one year to become knowledgeable combat officers and to be capable of serving as a regimental commander or similar. Here the training had to be organized according to military branch.

Learning Russian was compulsory in all three branches of the Academy. Learning English or another language was only allowed for those who could already speak and write Russian fluently. The National Defense Academy was to be set up by its commander, Major General Béla Király. The new institution began to operate on September 25, 1950, with personnel, weapons, materials and other equipment from the Staff Course and the Higher Commander Course.

The first academic year lasted from November 1, 1950, until September 23, 1951.

Thereafter, the measure accurately followed the proposals made in Major General Király's submission, except that Branch I students would also be selected by the head of the Chief Section of

Personnel, from among younger cadres "worthy and suitable in political and military respects." It emerges from this that such candidates were excused the entrance examination process. Students had to arrive by 12 noon on October 29, having taken at least two weeks of their annual leave.

In line with the measure described above, the commander of the Academy made a submission on administrative and personnel issues to the minister of defense on September 20, 1950.[7] Most of the instructors and administrative personnel were chosen from officers on the Higher Commander Course and the Staff Course. 14 officers, most of whom had served also in the National Defense Force of the Kingdom of Hungary, were not required. This category, which accounted for the bulk of the instructors, consisted mainly of officers who had been brought out of retirement during the forced expansion of the armed forces in 1948–1949. According to a specialist who served in the Academy at that time,[8] these officers were well trained, but ignorant of the Soviet principles, which they had to learn "on the job," hardly keeping ahead of their students. But more serious was the political uncertainty surrounding them.[9]

On October 18, 1950, the commander of the HAKAD submitted a document to the minister of defense entitled "Situation Report on the Work of Establishing the HAKAD."[10] In this it was reported that the Academy administration was running and only 5–7 extra personnel were required. The transfer of students was thought to have gone smoothly: 60 officers had arrived for Branch I, 25 for Branch II and 98 for Branch III. That meant 18 vacant places for I and 5 for II, but 5 officers too many for III.

The draft training syllabus was discussed with Soviet advisors and heavily revised on that basis, so that the version to be published on October 21 would "match the system and subject matter of the Soviet military academies to the fullest extent." The teaching staff had been doing three weeks of preparatory work since October 9, so that all the lectures, group discussions and introductory materials for November and December were either available in printed form or ready for the press.[11]

The "Training Calendar" consisted of the following:
1. *For the two-year Academy Course:*
November: "Techminimum" (technical minimum, that is, general knowledge for army officers—70 hours); company combat training (60 hours);
December–January: Battalion combat training (100 hours);
February–March: Regimental combat training (180 hours);
June–July: Fundamentals of divisional combat training.
2. *For the one-year Higher Commander Course:*
November: "Techminimum" (100 hours); company/battalion combat training (35 hours);
December–January: Regimental combat training (100 hours);
February–May: Divisional combat training (225 hours);
June–first half of July: Army corps combat training (100 hours);
Second half of July: Field army combat training (18 hours).
3. *For the Regimental Commander Course:*
November: "Techminimum" (100 hours); squad/company combat training (60 hours);
December–January: Battalion combat training (95 hours);
February–May: Regimental combat training (190 hours);
June–July: Divisional combat training and army corps information (80 hours).
For all three courses of training:
First half of August: Tactical travel, live shooting and map reading;
Second half of August: Final examination;
September 1–20: Division and army corps exercises;
September 21–23: End of academic year.[12]

The complement of the National Defense Academy on October 30, 1950 (regular/effective), showed the following picture: 2/3 generals (1 a student); 395/334 officers (223 of them students);

91/100 subalterns and corporals; 101/99 privates; and 197/149 civilian employees, making a total of 786/685.

Mounting Cold War hysteria appeared also in some figures in the October "Report on the Disciplinary Position."[13] During earlier courses, there had only been occasional "minor infringements of vigilance"; now two majors each received 13 days' confinement to quarters and a lieutenant colonel 15 days for such "offenses against vigilance." Efforts to "cleanse" the army may also have contributed to the fact that one colonel, four lieutenant colonels and two majors were pensioned off on November 1. Their high rank suggests that they had probably been officers of the Royal Hungarian Army as well.[14]

By December 1,[15] the (regular/effective) complement was 2/3 generals (1 a student), 395/363 officers (an increase of 29 in one month, of whom 248—25 more—were students, 20 on the Higher Commander Course, 78 on the one-year course and 150 on the Regimental Commander Course) including 12 from the State Security Authority (ÁVH), 92/103 subalterns and corporals (an increase of 1/3, 30 being students), 101/96 privates (no increase/decrease of 3); and 197/184 civil employees (no increase/increase of 35), making a total of 787/751 (increases of 1/66) serving in the institution.

Every month from November 2, 1950, onwards, a "Summary Situation Report" on the previous month was prepared for the minister of defense.[16] The one on October reported that the Academy was short of 27 officers, 10 subalterns, 11 corporals, 10 privates and 10 civilian employees. Teaching was affected: "Each teacher faces such a heavy task in preparing lectures and with the content of those lectures that it must sooner or later be at the expense of high-quality preparation for the work and thereby of the high-quality knowledge of the students." At the same time, it was pointed out that the shortage could not be solved by using outside teaching staff, as the frequent substitutes were often uninformed about the teaching as a whole and about the students.

Most of the students completed their tasks with great enthusiasm, but they coped with great difficulty with the heavy intellectual demands made on them from 5:45 in the morning to 10:30 at night. To make the material easier to understand, the faculties manned a consultation service throughout the periods of self-instruction, and the teacher in charge of a class would explain methods of taking notes. Some students had received only a low level of preliminary training (such as a course for reserve officers) and the study groups were chosen so that the better-prepared students could help the weaker ones, or the heads of classes and other teachers would pay special attention to the latter.

The disciplinary situation was thought to be generally satisfactory, but much more was expected of the students in their military conduct, vigilance and internal order. Penalties for lateness and lack of punctuality were imposed. The political situation and morale were rated as being generally good, as "both students and teachers have understood the importance of the struggle for peace—including the development and strengthening of the People's Army and mastery of Stalinist military science."

The National Defense Academy received a further important assignment during those weeks. Implementation of this was the subject of a "Plan for Military Training of Leading Party Functionaries,"[17] submitted to the minister of defense on December 18, and incorporated into the monthly timetables for January and February 1951. The course was planned to include 165 hours of instruction and 3 hours a week at the Academy, totaling 129 hours, over 10 months (January 2–November 1, 1951). The syllabus covered tactics (117 hours, 70 percent), staff service (14 hours, 14 percent), military-technical information on branches (18 hours, 11 percent), topography (6 hours, 4 percent), army administration (4 hours, 2 percent), and knowledge of imperialist armed forces (6 hours, 4 percent). Essentially, it was thought that leading Party functionaries should have knowledge at regimental or divisional level. A proposal was sent to the Academy, in a letter signed by

Colonel General Farkas,[18] calling for commanders' extension training for leading ÁVH officers as well.

Soon afterwards, the Ministry of Defense sent the Academy a document on "Party Functionaries and Functionaries of State and Mass Organizations Accepted for Military Training,"[19] which shows how seriously those in charge were taking the course. The 66 people attending included the following: Károly Kiss, president of the Party Central Committee; András Hegedűs, a secretary of the Party Central Committee; László Földes, head of the Central Committee's Cadres Department; Mihály Komócsin, deputy head of the Central Committee's Agitprop Department; János Csergő, state secretary at the Ministry of Heavy Industry; Ferenc Erdei, minister of agriculture; Zoltán Vas, head of the Planning Office; József Köböl, general secretary of the Construction Workers' Union; János Bruttyó, deputy general secretary of the Construction Workers' Union; Lajos Fehér, editor of the rural weekly newspaper *Szabad Föld*; Party secretaries and chief executives in districts and large factories.

The "Summary Situation Report" of December 1950[20] stated that the Academy still suffered from a shortage of 31 officers, 10 subalterns, 9 corporals, 13 privates and 6 civilian employees, and 80 percent of the unfilled places were teaching posts. The prescribed training assignments had been fulfilled and student discipline remained good. The November study results were also included in the report. The average grades for the two classes of Branch II (the Higher Commander Course) were 3.387 and 3.225 out of 5, the highest average being 4.285 (Major General Sándor Házi) and the lowest 2.285. The class averages in Branch I (the two-year Academy Course) lay between 3.666 and 3.4, with extremes of 4.61 and 2.00. In Branch III (the Regimental Commander Course), the best class average came from the infantry (3.844) and the worst from the artillery (2.64), while individual averages ranged between 4.6 and 2.

The branches had been able to keep two months ahead in preparing the lectures. It was reported to be important for quality that further Soviet advisors be sent to the Academy. It had been possible to clarify several questions of principle and technicalities during the officer methodology drills held by the two existing advisors, and this had helped to improve the quality of the training given. It was also hoped that Tactical and Military History "cabinets" for viewing teaching materials could be completed by the end of the year. Some deterioration in discipline had been found: for instance, an officer had wanted to take off the premises some material marked "Secret."

The "Summary Situation Report" for January[21] contains important information on events in the first month of 1951. It begins by referring to another document, issued on the previous day, which "features proposals the implementation of which is urgently necessary; please deal with my submission out of turn, or possibly place it on the agenda of the *Kollégium* of the Ministry of Defense, because the proposals affect the provinces of almost all the chief sections of the Ministry." There had been no material change in the situation of personnel. Only Branch III (the Regimental Commander Course) was able to meet its training target; the other two were nine days late due to a command exercise. The results of the "techminimum" examinations taken in the early days of January were as follows:

Examination Results in the "Techminimum"
(General Military Knowledge), January 1951

Branch	"Excellent"(5)	"Good"(4)	"Satisfactory"(3)	"Unsatisfactory"(2)
I. (Two-Year Academy Course)	1 (1.3 %)	30 (38.9 %)	34 (41.2 %)	12 (15.5 %)
II. (Higher Commander Course)	-	13 (33.3 %)	18 (46.2 %)	8 (20.5 %)
III. (Regimental Commander Course)	-	52 (33.3 %)	77 (49.4 %)	27 (17.3 %)

Two students each in Branches I and III had not taken the exams, due to hospital treatment. A role in the results had also been played by the fact that the students "come in voluntarily on Saturday and Sunday as well... the teachers are constantly at the students' disposal for consultation purposes, in many cases even on Sunday." The students with a "not satisfactory" grade would retake the examinations in February.

There had been further deterioration in discipline in January. Several people were disciplined for shortcomings in vigilance (for example, the classroom had been locked but not the secret box). "We will overcome the existing shortcomings in confidentiality and vigilance with frequent surprise checks and enhanced political education," the commander of the Academy and its political officer promised.

Turning to the political situation and morale, the report said that students had shown great interest in the events of the Korean War, Eisenhower's European tour, and the Yugoslav military exercise. But above all, "Everyone greatly looks forward to the Second Congress of the Hungarian Workers' Party. We know what the Party's leading role means. We expect the Congress to disclose the existing errors and chart the path for the future. We are preparing to celebrate this great event for our country and Party with pledges and good work from our staff."

Mention has already been made of another document, issued on January 31. This was a very thorough 12-page piece by Major General Béla Király and Colonel György Sulyán, sent to their superiors. Entitled "Submission on the Subject of Resolving Basic Questions concerning the HAKAD,"[22] even its first lines were unusually tough: "I report that the National Defense Academy began the academic year in a spirit of improvisation on the most essential questions. So the work of organization was not decided upon in good time, timely action was not taken to ensure the cadre requirements, and finally the construction work needed for the

1950–1951 training year was based on incorrect calculations of scale and quality." Many other difficulties were also mentioned as having jeopardized the quality of the work done. Most of these could have been overcome if the work of organization had begun in time. "I therefore make in the following memorandum some basic proposals that will create the requisite conditions for the 1951–1952 training year and to some extent improve the work in the current 1950–1951 training year as well," the Academy Command declared.

I/A. Proposals
Connected with the Tasks of the 1951–1952 Academic Year

This section of the submission suggested that Branch I should continue with its second year. Branch II—as of that year—should begin its one-year studies in four groups (infantry, armored, artillery and air force classes) and finish its training in the autumn of 1953. Branch III should begin the 1951–1952 academic year with the four branch classes mentioned above, but it should be examined whether there was not a need for training of engineering and signals troops officers as well. If that was decided, it would be possible to introduce second and third branch groups with ten officers each in the classes for engineers and for signals troops. That, however, would call for some extra teaching staff.

I/B. Proposals for the Reorganization of the Present Branch II
(Higher Commander Course)

It was proposed that this should be made a year's course instead of one lasting five months. Five months had sufficed for the first course, as "the students were excellent cadres in every respect," capable of preparing to perform the functions of higher commanders in that time. But the present students had not been selected so rigorously, and there would be some who were incapable of reaching the training target in such a time. The same would probably apply in the future, and it was proposed that the course should last one year even in the 1951–1952 training year.

It would be worth starting a third (teacher-training) class in this group, the graduates from which "would be capable of being assistant instructors at the HAKAD and becoming instructors in tactics for the cadet schools." Although they would essentially study the syllabus of the Higher Commander Course, training methodology would be the "major subject" for them. This would make it possible to provide training for "individual students" at the National Defense Academy, for, in line with current practice, there would also be "tasks to accomplish in the coming training year, such as providing individual training for leading Party cadres, political officials in higher posts, and so on."

I/C. Branch III (One-Year Regimental Commander Course)

The only change proposed was that only 10 instead of 20 air force students be trained, so that the course would begin with 140 students.

In all, the 1951–1952 academic year was expected to begin with the following numbers:

Branch I:	240
Branch II:	60
Branch III:	140
Individual students:	80
Total:	*520*

II. Organization

It had been found that the teaching staff could meet the quantitative and qualitative requirements only at the expense of very great efforts, and so it would be to the purpose for the present strength of the National Defense Academy—discounting students: 87 officers, 29 subalterns, 67 privates and 17 civilian employees—to be raised. There was a ratio of 1 student to 1.8 personnel (officers, men and civilians) and 0.5 officers at that time. "The decision taken in principle by the Comrade Defense Minister for the 1950–1951 year was that this proportion should not be reached even in the 1951–1952 year," said the report.

III. The Position with the Teaching Staff

On the one hand, the staff numbers were inadequate (only 51 of the 74 teaching posts had been filled, making a 30 percent shortfall), and, on the other, these did not meet the requirements of the Academy at present, let alone in the future. "Of the existing 51 teachers, 36 were officers in the Horthy army, so that 70.6 percent of the teaching cadres are old officers. On the foundation of the National Defense Academy and since then, we have received out of the cadres trained in the Soviet Union or graduating from the Higher Commander Course or Staff Course only 15 teachers, who make up 29.4 percent of the teaching cadres." The position was exacerbated further by the presence on the teaching staff of 8–10 persons "whose retention at the National Defense Academy is undesirable."

To fill the gaps, it was proposed that the Academy should receive regularly from the Soviet training institutions an intake that would provide at least one foreign-trained staff member per faculty, and, within that, one for each "special subject." It was calculated that at least ten such persons would be needed by the end of the training year. In addition, the Academy wished to retain twenty of those currently on the Higher Commander Course.

Finally, the document repeated the proposal under Head II that teacher training should commence on the Higher Commander Course and continue "until the teacher requirements of the National

Defense Academy have been met out of students who have completed the two or three years. When that time comes, only those who were trained in the Soviet Union or who have successfully completed the two- or three-year branch at the National Defense Academy will be allowed to teach there."

IV. Organization of Spheres of Authority over Commendation and Discipline

The following recommendations were made:
- The authority of an army corps commander laid down in the Service Regulations for the commander of the Academy is realistic;
- The studies and administrative deputies to the commander of the Academy and the branch commanders should have the authority of a divisional commander;
- The faculty and department—political, training and logistics—commanders and heads of sub-departments and departments should have the authority of a regimental commander;
- Student class commanders should have the authority to give simple and strict rebukes in words and verbal commendations to all subordinates.

V. Proposal for Establishing a *Kollégium* of the National Defense Academy

It was thought necessary to establish a *kollégium* of the heads of the institution, in the light of the study being done at the Academy, the research that would soon begin, and the increasing size of the organization. The tasks of the *kollégium* would be as follows:
- To ensure that Soviet military studies were mastered correctly and the Academy's experience in doing so was evaluated and put to use;

- To regulate the scientific work in the Academy and perfect
  its methods of training;
- To put proposals based on the experience obtained to the
  chief of staff.

The *kollégium* would include as *ex officio* members the commander and political officer of the Academy and the head of the
Political Department. The other members would be appointed by
the chief of General Staff. It would start operating on May 1, 1951.

VI. Selection of Students

The January examinations in the "techminimum" had shown that it
was not certain in the case of 13–20 percent of the students that
they would complete their studies successfully. This proportion
could rise in future with the increase in the complement of students.
Another problem was that if the period of training increased to
three years, the future dropouts would be "dragged along" for an
even longer period, which was extremely uneconomic. "The situation might be changed by having us select students through an
entrance examination even in the 1951–1952 training year."
    The specific proposals in this regard were as follows:
- Students for the one-year courses (for regimental commanders and higher commanders) would continue to be directed
  to the Academy by the Chief Section of Personnel of the
  Ministry of Defense.
- Students would be admitted to the first year of the three- and
  two-year branches only after an entrance examination. This
  would cover both basic military knowledge (topography, battalion tactics and the "techminimum") and general knowledge (basic arithmetical calculations and knowledge of the
  Hungarian language). Knowledge of the Russian language
  would be an advantage.

- The students would be chosen with two or three applicants per place and the principle of voluntary application introduced. "At the same time as applications are invited, political work should be initiated to ensure that the very best cadres apply," ran the proposal.

VII. Other Matters

A. *Grading of Individuals*

Students of the Academy could receive the following grades:

"5" or "excellent" was the grade to be awarded to those students who mastered the training material perfectly and could carry it out in practice, who could successfully organize coordination among branches, who could perform the practical tasks without help from superiors or teachers, who stood out from their peers in commanding ability, and who emphasized self-training;

"4" or "good" was the grade to be awarded to those students with a firm grasp of the syllabus, able to apply their knowledge correctly in practice, arguing correctly for their decisions; they would solve their tasks with minimal cooperation from their superiors or teachers and display above-average ability as commanders;

"3" or "satisfactory" was the grade to be awarded to those students who had learnt the material adequately and could present it correctly, but occasionally inaccurately; tasks were generally solved with some teacher assistance;

"2" or "not satisfactory" was the grade to be awarded to those students who had only superficial knowledge of the essence of the material, who could not apply the theoretical material in practice, and who could not solve the tasks even with teacher assistance.

Apart from the criteria mentioned, attention had to be paid also to determination, practical common sense, and expression in verbal, written and chart form. (The three criteria are given here in order of importance.)

B. Introduction of Multiplication Numbers

To give an emphasis to the most important subjects when assessing
students' study results, the Academy would introduce multiplication numbers. Attention should be paid not only to importance, but
to the number of hours of tuition received. Although the minister of
defense in his order had classed "logistics" as an auxiliary subject,
the Academy Command recommended that it should become a
principal subject, since "The training at the National Defense
Academy concerns units where the organization and securement
of logistical services is an extremely important task for the commander."

C. Definition of the Individual's Place in the Ranking List

This positive intention was defined in the following somewhat
complicated way.

A list was prepared to show multiplication numbers—reflecting the importance of the subject—with which the grade given by
the instructor had to be multiplied.

The thus defined grades had to be added up.

This total had to be divided by the sum total of the multiplication numbers.

The thus attained grades, 2 though 5, had to be calculated up
to two decimal points. That determined the student's overall
achievement in his study, which was the basis of his place in the
ranking list.

"5" or "excellent" meant that the average calculated by the
method just described was a minimum of 4.8 and did not include
any grade of 3 or lower in general tactics, the tactics of the student's
own branch, or a political subject (Marxism-Leninism, history or
political economy);

"4" or "good" meant a student whose average was at least 3.7
and included no grade of 2 or less in the subjects just listed;

"3" or "satisfactory" meant that the student's average was no worse than 3.

These efforts to improve the quality of a correct assessment of the situation by the Academy Command testified that the Command had assessed the situation correctly. Further evidence that they chimed in with ideas at a higher level in the hierarchy came shortly afterwards, when a Council of Ministers' order on further development of the armed forces appeared on February 17, 1951.[23] In the "Summary Situation Report" for February 1951,[24] the commanding office reported that Branch II (the Higher Commander Course) had been on a command exercise involving a divisional attack. The timetable assigned 18 hours for this, "but the students, as a pledge for the [Communist Party] Congress, had requested that the exercise be set for Saturday and Sunday, so that they could absorb the material perfectly and thereby enhance their knowledge. This command exercise, by showing how the theoretical material covered so far could be put into practice, has raised the military preparedness of the students considerably."

The (regular/effective) complement of the National Defense Academy on March 1, 1951, was as follows: 2/3 generals, 398/380 officers, 92/84 subalterns and corporals, and 205/210 civilian employees, making a total of 798/792 people and 16/16 horses.[25] So there was a very slow, modest increase in strength to be seen.

At this time, the Academy roll[26] was submitted to the Chief Section of Personnel of the Ministry of Defense. Among the names were several who would later hold high leadership posts in the army: Major General Béla Király, commander of the Academy; Colonel György Sulyán, political officer of the Academy; Lieutenant Colonel Lajos Tóth, head of the Training Sub-department and deputy head of studies (later acting chief of General Staff, and commander of the Miklós Zrínyi Military Academy [ZMKA]); Colonel Imre Lóránt, chief artillery instructor (later second-in-command of the ZMKA, the first to hold a candidacy of the Hungarian

Academy of Sciences in military studies and then the first doctor of military studies); Lieutenant Colonel Sándor Zsilinszky, head of the Military History Department; Major General Béla Kerekes, commander of the Regimental Commander Course (later founder commander of the Ferenc Rákóczi II Military Secondary School).

"Soviet military advisors began to assist in 1951 with theoretical and practical training. The chief advisor at the Academy was Major General Samsonov, while teaching in all-branch tactics was assisted by Guards Colonel Voloshin."[27] To make cooperation with advisors more effective, the commander called for "urgent transfer to the National Defense Academy of an adequate number of competent interpreters."

The disciplinary situation still fell short of what could be expected of a National Defense Academy. Two reasons were given: the unsatisfactory "social composition and spirit" of the teaching staff (mentioned several times), and the "quality" of the political officers. By the latter, the author of the report meant that most of the political officers were young and inexperienced, and some had only a low level of ideological training, so that ideological errors would creep into their teaching and they could not give ideological support for the teachers' work. The draconian measures taken to improve discipline were exemplified by a case that month of a lieutenant colonel sent before the military prosecution service for burning a secret document by mistake.

"Finally, I report that behind the grave difficulties apparent at the National Defense Academy since its foundation (bad working conditions, tardy improvement in discipline, very many offenses against vigilance) there can be seen the hand of the enemy. Tangible evidence of this now emerging is that on April 2, two enemy leaflets were found in the corridor of the Command building."

On April 13, 1951, the Academy Command presented a specific administrative proposal to the minister of defense.[28] This suggested a ratio of 1:1.8 between students and teaching staff.

The Proposed Organization of the Honvéd (National Defense) Academy (HAKAD)

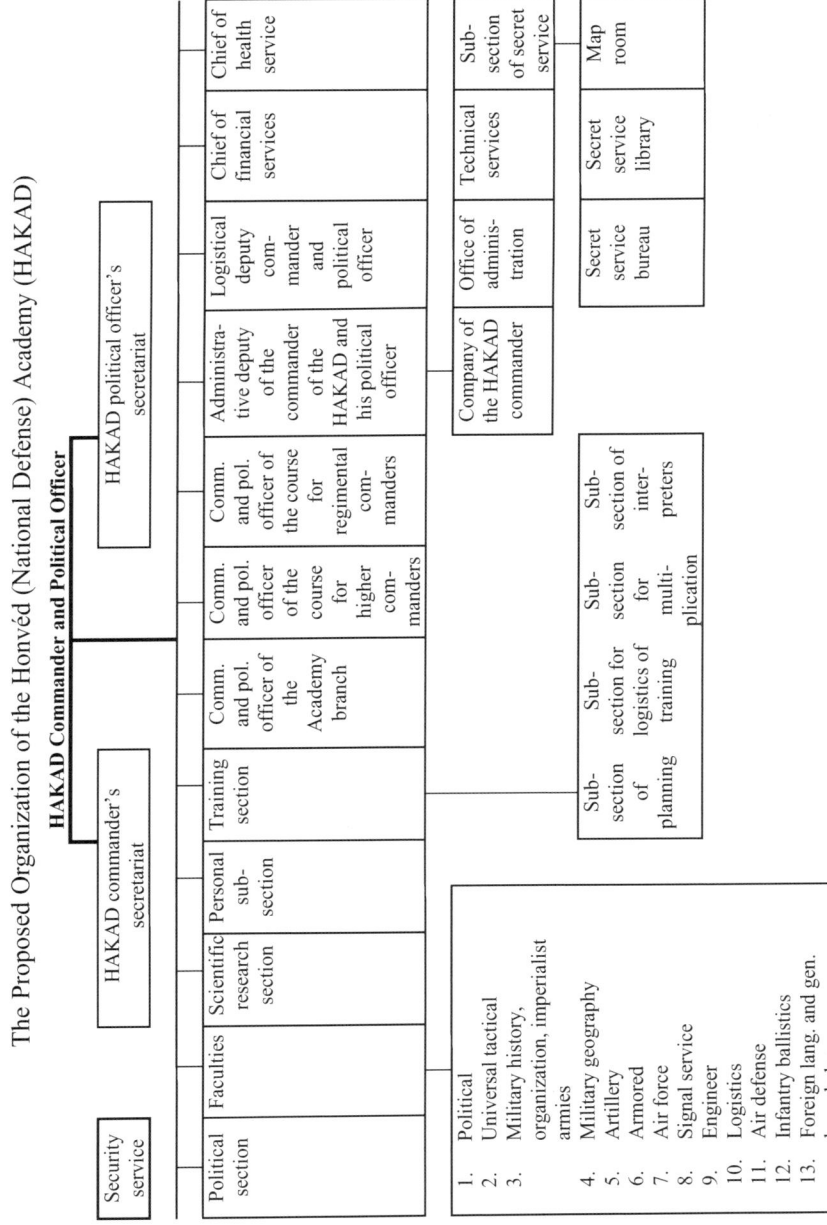

HAKAD Commander and Political Officer

| HAKAD commander's secretariat | HAKAD political officer's secretariat |

Security service

| Political section | Faculties | Scientific research section | Personal sub-section | Training section | Comm. and pol. officer of the Academy branch | Comm. and pol. officer of the course for higher commanders | Comm. and pol. officer of the course for regimental commanders | Administrative deputy of the commander of the HAKAD and his political officer | Logistical deputy commander and political officer | Chief of financial services | Chief of health service |

1. Political
2. Universal tactical
3. Military history, organization, imperialist armies
4. Military geography
5. Artillery
6. Armored
7. Air force
8. Signal service
9. Engineer
10. Logistics
11. Air defense
12. Infantry ballistics
13. Foreign lang. and gen. knowledge

| Sub-section of planning | Sub-section for logistics of training | Sub-section for multi-plication | Sub-section of inter-preters |

| Company of the HAKAD commander |

| Office of adminis-tration | Technical services | Sub-section of secret service |

| Secret service bureau | Secret service library | Map room |

The reference a few days earlier to the Soviet advisory working plan bred results. On April 28, 1951, Minister of Defense Farkas issued an order concerning the tasks of the National Defense Academy in the 1951–1952 training year,[29] prescribing that training should continue in the second academic year, for those from the infantry, artillery, armored and air force branches who had successfully completed the first year. New first-year courses would begin in the infantry, artillery, armored, air force, anti-aircraft artillery, signals troops and engineering branches. The one-year Regimental Commander Course should start; within it infantry (border guard) artillery, armored, anti-aircraft artillery, air force, engineering and signals troops subjects should be included. The Higher Commander Course should also start, but only in an all-branch version. The period would be nine months, with 60 students, the goal being to meet the tactics-tuition needs of the National Defense Academy and the cadet schools from those completing it. Both the increase in student numbers and the idea for augmenting the teaching staff were accepted. But quality development was postponed: the authorities decided that "three-year training would not begin" in the 1951–1952 training year.

Also set were the numbers of students within each branch:

Student Numbers by Type of Formation, 1951–1952 Training Year[30]

Type of formation	First Academy Year	Second Academy Year	Regimental Commander Course
Infantry	19	30	40
Armored	15	15	20
Artillery	25	15	20
Air Force	20	10	10
Anti-aircraft Artillery	-	10	20
Engineers	-	10	15
Signals Troops	-	10	10
River Fleet	-	-	1
Chemical Defense	-	-	1
Logistics	-	-	1
Total	*79*	*100*	*138*

To allow the National Defense Academy to prepare in an organized way for the coming academic year, the minister ordered that the new table of organization and personnel should be issued by the Ministry by May 15. The new table of personnel of the Academy had to be filled gradually, so that every position was filled by September 1, 1951. All candidates for the Academy in 1951–1952 were to take an entrance examination. For the entrance examinations, there had to be twice as many candidates from which to choose, and they were to be held continually from July 15 to August 15.

A. Training in the *two-year Academy branch* was intended to produce officers with good all-branch knowledge. Their military expertise should suit them for posts as regimental commander or divisional chief of staff, or in staff of army corps or central organizations. The emphasis also in the specialist branches was to be on general tactical training, but students would have to have good specialist knowledge of their own branch as well. They would need to know the tactical principles for regiments and divisions, receive guidance on the command of an army corps, and gain general knowledge of the army operations.

The training assignments were intended to enable students to gain familiarity with organizing, preparing and securing modern mixed-branch combat, to practice commanding these branches in combat, and to develop theoretical and methodological abilities in training troops.

The result for the students would be that they would learn the basics of Marxist-Leninist theory and become clear about current events, learn the basic principles and use of branches in modern, mixed-branch warfare, and organize and command cooperation among these, as well as learning the method of reaching a decision. They would become familiar with the staff work in various types of combat. They would learn the theoretical bases of ballistics, so that they could direct the fire of infantry or of weapons of their own branch during combat. They would master the material of their

branches and its use in battle. They would be capable of conducting the combat and political training of their subordinates and
applying regulations, directives and orders in practice. They would
learn to read and write Russian to a level where they would know
the elementary specialist military expressions, so creating a basis
for "study of Stalinist military science in the original language." To
develop their general education, they would obtain knowledge of
Hungarian, arithmetic, military history, physics, geography and
military geography suited to their rank.

The training of the students was to take place in line with the
requirements of the regulations and directives: "We will utilize to
the full in training the experiences of the Soviet Army gained in the
Great Patriotic War." The students were to be trained in a spirit of
high-level, self-aware military discipline, so that they would make
similar demands on themselves and their subordinates.

B. Training in the *Regimental Commander Course* was intended to turn the students into well-informed field officers by the fall
of 1952. Based on their general military knowledge and expertise
in their own branch, they were supposed to be capable of filling the
posts of regimental commander and regimental chief of staff or of
holding equivalent central posts.

The training tasks were intended to enable students to obtain
the knowledge of their branch required to command a regiment or
independent battalion in combat with modern, mixed-branch higher units. They were to gain experience in commanding such forces.
Students were to develop their theoretical and methodological
capabilities in training lower units. The results of the course were
described in essentially similar terms, with the additional requirement of general education. This applied to the Academy branch and
to the other two courses.

C. The *Higher Commander Course* was intended to make students competent to hold a divisional command post, or to become
the political officer or chief of staff of an army corps or of a division.

The tasks of the training were expressed word for word in the same way as those of the Academy branch. Neither general education nor Russian-language requirements were mentioned.[31]

The Chief Section of Personnel of the Ministry of Defense issued an order entitled "Provision on Application for the Entrance Examination to the National Defense Academy,"[32] which can be summarized as follows:

Commanders were to call upon all officers to apply. Those eligible would be those who had completed a branch school or the Kossuth Academy in 1950 with a grade of "excellent" or "good," and have at least one year's field service behind them, and all other first-rate officers able to complete the Academy Course successfully, based on outstanding military and political capabilities and good organizing skill. In addition, commanders of higher units and the requisite persons in Ministry of Defense organizations were to submit for the nine-month Higher Commander Course suitable officers who had served in diligence, regimental chief of staff or similar central posts or even higher posts. There was a general requirement of complete political reliability, moral fiber, rapid understanding of the position of regiment or independent battalion or higher commander, and the requisite physique. Candidates were examined in Marxism-Leninism, general tactics, branch tactics and technical data, topography, basic arithmetic and the Hungarian language. They might also take an optional examination in the Russian language, which would be an advantage where scores were otherwise the same. The candidates for the entrance examination were to be sent material that would allow them to prepare for it.

On April 13, the Academy Command sent its superiors a new administrative proposal, to which the response was an order on May 25, 1951, concerning new administrative officer posts,[33] in which Lieutenant General István Szabó, deputy minister and head of the Chief Section of Personnel of the Ministry of Defense, posted or confirmed a large number of officers who served in the HAKAD, with effect from June 1, 1951 (when the new table of organization would come into force). This raised the teacher short-age to 80, with places for a total of 89 officers, 24 subalterns and 55 appointed and 9 contracted civilian employees still unfilled. The shortage of translators and Russian-language typists was mentioned particularly.

The new Higher Commander Course began on May 3, 1951. Its members had taken a written examination for information pur-poses. The examination on basic branch knowledge revealed that 80 percent of students were ill-informed about general tactics and able to answer the questions only in general terms if at all. With staff service, the experience was the same, with the addition that several students now had to make a working map for the first time. There was a poor level of artillery knowledge in 70 percent of cases, while in armored and air force knowledge, the "ideas" were said to be very patchy and confused. The most acceptable levels of information were found in engineering and topography. Students did not know about the combat procedures of the imperialist armies, and even in their own army their knowledge tended to be confined to the old organization.

The training situation of the *special classes*

The *Party functionaries* had completed their study of defense of an infantry regiment that month, and a basic lecture on the defense of infantry divisions had been held. They were now receiv-ing lectures on staff service at infantry division- and infantry corps-level, in which the students were said to be showing great interest.

The training of the *ÁVH officers'* group was proceeding as planned, but as student absenteeism of 50–60 percent was causing serious problems, the Academy Command proposed issuing an order that made participation in the activities compulsory.

The group of *generals and higher-ranking officers of the Political Chief Section of the Ministry* showed the greatest delay (one month) as "the majority of the comrades generally attend the activities irregularly." Cases of 50 percent absenteeism had occurred. If this practice persisted in the summer months, the Academy Command would propose that the students be examined in the syllabus completed so far and continue the course in the fall. A decision was requested on this.

Based on the new table of organization of the HAKAD, (regular/effective) personnel on July 1, 1951,[34] comprised 5/2 generals, 623/419 officers (255 of them students), 81/61 subalterns, 124/138 privates and 209/177 civilian employees, making a total of 1,042/797.

A former course student, Major General Sándor Házi, commander of the air force, requested from the Academy on July 2, 1951, "the institution of a field officer course."[35] He stated in his letter that the minister of defense had approved his proposal, since "the rate of development in the autumn order of battle requires that we should train new air force field officers." The goal of the course was to train officers in the company, wing and aircraft engineering detachment categories, and to train officers capable of organizing and directing combat training in schools and units, the combat tasks of a sub-unit and wing, material and technical logistics of the troops, and carrying out inspections.

The 50 students were to be selected by the Chief Section of Personnel of the Ministry of Defense from the reserve officers being transferred into the regular force, making sure that those most suited politically and professionally were chosen for the course.

According to the June "Summary Situation Report" of the National Defense Academy,[36] there was a shortfall of 139 officers (of which 70—42.2 percent—were teachers), and 92 appointed and 80 contracted civilian employees. All this seemed to pale beside the problems in the Air Force Faculty, where the shortages "place at risk the attainment of the training goals set for the air force students, and it would additionally affect air-related training elsewhere in the HAKAD. For the experiences that the currently present Soviet advisor comrades can offer in the HAKAD cannot be taken over by the Air Force Faculty, and this valuable assistance cannot be utilized in the future. The gravity of the situation is increased because there is no air force student in the Higher Commander Course class..."

The Air Force Staff Service Course started on July 10, but the storm over the Air Force Faculty did not blow over. Despite their moral and political shortcomings, most of the students were found to be capable of development, although the overwhelming majority of them were not capable of filling a responsible post. "However, they can produce valuable work under the hand of a strong commander. The impossible situation in this branch will be resolved—as a key assignment for the HAKAD—but in order to avert similar events, it was requested that the impossible cadre situation in the Air Force Faculty be resolved urgently," ran the repeated Academy call.

The situation in other areas was not so critical. As far as the training situation went, the work was continuing in all three branches as planned. The entrance examinations for the 1951–1952 year went smoothly and the officers called met the prescribed requirements to a middling degree. The Chief Section of Personnel of the Ministry of Defense did not call in any officers from the air force for the examination. The training of ÁVH officers also went according to plan, and the number of absentees steadily declined. Production of the lectures and assignments for the new training year went ahead; the plan was to complete 50 percent of it by September 30.

The nerves of those at the Academy were also jarred by another incident, although few people yet realized that it was a further instance of the witch-hunts spreading through the People's Army and the Academy as well. Many people were astounded when a warrant for the arrest of Lieutenant Colonel Sándor Zsilinszky,[37] head of the Military History Faculty and among the best-known teachers at the Academy, was issued by the Budapest Central Military Prosecution Office, alleging the following: "In Budapest, during his permitted leave between July 2 and July 21, 1951, he left his post between July 13 and 17 for the purpose of withdrawing himself permanently from every kind of military service duty, and remains absent to this day (the crime of desertion breaching Military Penal Code §40)." The Office also instructed the Academy Command as follows: "if the accused appears, please escort him to the prison of the Budapest Central Military Prosecution Office (Budapest II, Fő utca 70–78)." Yet his comrades at the time knew that Lieutenant Colonel Zsilinszky was already under arrest.[38]

But this was only a forewarning of the "game" that was to be played. The first move came on August 17, 1951, when Colonel General Mihály Farkas, minister of defense, relieved Major General Béla Király of his post with immediate effect and placed him in the unpaid group of reserve personnel.[39]

At the same time, Farkas appointed Major General Béla Kerekes as acting commander of the National Defense Academy. Naturally, there is no word in this order of an arrest. According to an old colleague, the establishment of the National Defense Academy and its successes in its first academic year were "to the merit of Major General Béla Király, who was decisive, very knowledgeable and considered, yet capable, by making swift decisions, of carrying out the complicated, constantly changing system of tasks without major mistakes."[40]

The "Summary Situation Report" for August was already signed by Kerekes and Captain Béla Kránitz, political officer of the National Defense Academy. But the question that most occupied the staff of the Academy could not be avoided. It was addressed in Point 1: "Due to the change of commander and political officer at the HAKAD, some feeling of uncertainty arose in some of the students and the teachers, although nobody discussed the matter openly. Some guesswork was discernible, but the majority continued to do their duty calmly; indeed a greater work momentum could be seen, especially among those in leading posts. The students, at an assembly, expressed a desire that the reasons for the change of commander and political officer be made known to them."[41]

On the training situation, the report noted that the Higher Commander Course had ended on August 18. The second semester's examinations, the live artillery exercise and the final examination had been held during the month, which was a big advance. The entrance examinations went smoothly. The weakest candidates were sent by the Armored Command. Nobody under the Air Force Command applied. The Party functionaries studied divisional defense, followed by battalion and regimental attacks. The state security course completed their study of regimental defense. Their tactical knowledge had improved, but their staff work was still judged to be weak. The high absenteeism had ceased.

According to the Academy "Summary Situation Report" for September 1951,[42] the most important training achievement of the month was the fact that in the final examinations for the Higher Commander Course and for the first year of the Academy, there were satisfying results for Marxism-Leninism: 87 percent of students on the former and 83 percent of those on the latter received a grade of 5 or 4. In the general tactics assessment of these two branches, there were substantial differences between the oral and the written tests in the proportion of students receiving a 5 or a 4 (64/43 percent and 73/41 percent respectively). Interestingly, the Russian-language examinations went better for the Higher Commander Course than for the Academy Course (62 and 42 percent respectively receiving a 5 or a 4). The combined study averages for the 1950–1951 year showed similar trends, with all three branches, individually and combined, showing more than 70 percent of the students receiving a 5 or a 4. The Higher Commander Course showed a particularly high proportion of 5s ("excellent").

Notes

1. László Csendes and Tibor Gellért, *Kronológia a Honvédség történetéből 1945–1990* [Chronology of the History of the *Honvédség*, 1945–1990] (Budapest, 1993), p. 91.
2. *Ibid.*
3. *Ibid.*, pp. 91–92.
4. Dr. Lajos Móricz, "A magasabb tisztképzés magyarországi történeté-nek elvi, szervezeti és vezetési kérdései" [Theoretical, Organization-al and Leadership Questions of the Hungarian History of Higher Officer Training], *Akadémiai Közlemények* 178 (1991): 16.
5. Honvédelmi Minisztérium, Hadtörténelmi Levéltár, Hadiakadémia [Ministry of Defense, Military History Archive, Military Academy] (hereafter cited as HM HL HAKAD) 01349/Mag. pk. tanf. [Higher Commandership Training Course]–1950 (Sept. 1).
6. HM HL HAKAD 01454/Tö. tanf. pság. [Command of the Staff Course]–1950 (Sept. 15).
7. *Ibid.*, 01480/Mag. pk. tanf.–1950 (Sept. 20).
8. Written statement by Lieutenant Colonel Géza Buzsáki (ret.), September 22, 2002 (in the author's possession), pp. 2–3.
9. HM HL HAKAD 1/P.–1950 (Sept. 25).
10. HM HL HAKAD 0244/H. Akad. P.–1950 (Oct. 18).
11. This entailed huge efforts, as huge quantities of teaching material had to be prepared for the three branches, before the students arrived, and then continually, keeping some weeks ahead of the instruction schedule. It often emerged after the aids had been prepared that they did not suit the requirements, partly because the Academy had ini-tially only one advisor, later two, and partly because the inexperience of the interpreters working with them became a source of frequent errors. See Buzsáki, *op. cit.*, p. 2.
12. HM HL HAKAD 0252/Techn. szolg. [Technical Services]–1950 (Oct. 18).
13. *Ibid.*, 0493/H.A.–1950 (Nov. 4).
14. The historian's hunch proved correct. After reading the manuscript of this book, a fellow officer related how Lieutenant Colonel Ákos Vizi-Makkay, one of those pensioned off, attended an officers' meet-ing already in civilian clothes, leaving his briefcase and personal

belongings in his office. Security men burst in on the meeting and tipped the briefcase onto the chairman's desk, producing secret documents, of course. Vizi-Makkay left in handcuffs. See Buzsáki, *op. cit.*, pp. 4–5.

15. HM HL HAKAD 0252/1.Techn. szolg. [Technical Services]–1950 (Dec. 1).
16. *Ibid.*, 0720/H. A.–1950 (Dec. 1).
17. *Ibid.*, 0952/H. A–1950 (Dec. 18).
18. *Ibid.*, 0982/H. A.–1950 (Dec. 24).
19. *Ibid.*, 0983/H. A. PK.–1950.
20. *Ibid.*, 0986/H. A.–1950 (Dec. 29).
21. *Ibid.*, 0228/H. A.–1951 (Feb. 1).
22. *Ibid.*, 0225/H. A.–1951 (Jan. 31).
23. Csendes and Gellért, *Kronológia a Honvédség történetéből 1945–1990*, p. 108.
24. HM HL HAKAD 0433/H. A.–1951 (March 5).
25. *Ibid.*, 0252/4./Techn. szolg.–1950 (Feb. 28, 1951).
26. *Ibid.*, 0537/H. A.–1951 (exact date unknown, but covering Oct. 1950–May 30, 1951).
27. Dr. Gyula Balogh, *A hadiakadémiai képzéstől a szervezett doktori (PhD) képzésig. Tanulmány* [From the Military Academy Training to the Organized Doctoral (PhD) Training. A Study] (Budapest, 2001), p. 10.
28. HM HL HAKAD 0850/H. A.–1951 (April 13).
29. HM HL HVK Titkárság [Secretariat]–1951/T. 5. doboz, Kik. ügyek [Training Affairs]: the research of Dr. Antal Oroszi, general (ret.).
30. 02312/HVKF.–1951 and 04565/HVK. Hdm. Csf. Kik. o.–Order No. 1951.
31. HM HL HAKAD 0982/H.A.–1951 (May 2).
32. *Ibid.*, Box 303: the research of Dr. Antal Oroszi, general (ret.).
33. *Ibid.*, 01318/H.A.–1951 (May 30): the research of Dr. Antal Oroszi, general (ret.).
34. *Ibid.*, 0252/6/Techn. szolg. [Technical Services]–1950 (June 30, 1951)
35. *Ibid.*, 01665/H.A.–1951 (July 6).
36. *Ibid.*, 01608/H.A.–1951 (July 3).
37. *Ibid.*, 2317/H.A.Szü.–1951 (Aug. 13).

38. According to Buzsáki, a close friend, Lieutenant General Zsilinszky was arrested in the street, near his mother's apartment. He was sentenced in the Béla Király trial to several years' imprisonment, but freed in the fall of 1956 and rehabilitated and promoted to general at the beginning of the 1990s. He died in 1993. See Buzsáki, *op. cit.*

39. HM HL 02446/HVKF–1951 (Aug. 18): the research of Dr. Antal Oroszi, general (ret.).

40. Buzsáki: *op. cit.*, p. 3. The arrests took place close to the time when the name of the armed forces was changed from *honvédség* (National Defense Force) to *néphadsereg* (People's Army): June 1, 1951. Csendes and Gellért, *Kronológia a Honvédség történetéből 1945–1990*, p. 112.

41. HM HL HAKAD 02182/H.A.–1951 (Aug. 31).

42. *Ibid.*, 02556/H.A.–1951 (Oct. 6).

Chapter VI

The National Defense Academy's Second Academic Year
(October 25, 1951–October 10, 1952)

The Cold War continued in the fall of 1951 and in 1952. The Korean War was still at its height. The United States signed a security treaty with Japan on August 8, 1951, and a military pact with the Philippines on August 30. Winston Churchill returned as Conservative British prime minister on November 3, 1951. On May 26, 1952, the United States signed a general treaty with West Germany, while the Pinay government in France signed the European Defense Community treaties on the following day. On October 3, 1952, Britain carried out its first atom bomb test on the Montebello Islands, which belonged to Australia. Under international conditions such as these, Hungary's political leaders had to prove that "Hungary is not a crack, but a strong bastion in the wall of Communism," as one oft-repeated slogan put it. So in parallel with that or consequent upon it, "the class struggle heightened" in Hungary too. There were feverish efforts to meet the norms of the First Five-Year Plan, which would turn the country into a "land of iron and steel." And the advances needed defending. A new generation of reliable officers was being trained up at a forced pace: 5,008 new officers were commissioned on September 16, 1951, in various institutions, and 4,020 students were admitted in the fall for first-year basic officer training. Yet there must have been concerns about whether that would be enough, as the officers' schools had dismissed 1,036 students altogether in 1950–1951: 708 for political reasons, 130 for moral shortcomings (such as drunkenness or theft), 61 after criminal convictions, 18 for unsuitable behavior, and 119 on medical grounds.[1]

It has been seen that a government order of June 1, 1951, changed the name of the army from National Defense Force (*Honvédség*) to Hungarian People's Army (*Magyar Néphadsereg*), and September 29 was declared People's Army Day.[2] Meanwhile the People's Army progressively became a Party army, and not just because the minister of defense on November 1, 1951, changed the official salutation from *bajtárs* ("comrade" in the military sense) to *elvtárs* ("comrade" in the political sense),[3] for the army command was changing rapidly as well. By this time, the Hungarian People's Army was *de facto* subordinate to the top Party leadership, not to Parliament, and the Party controlled all high military appointments. Among the consequences was an alteration in social composition: by the end of 1951, 53 percent of officers came from worker families, 24.5 percent from the peasantry, 18 percent from clerical strata, and 4.5 percent from other social backgrounds, while 71 percent of officers belonged to the Communist Party and only 9 percent had served in the old Royal Hungarian Army.[4]

Further development of the armed forces was arranged, although the plan to wage war against Tito—the real reason for the huge increase in the armed forces of the Soviet bloc—had already been cancelled. However, while Stalin was still alive this change of mind was not publicly announced. Thus the huge burden on the population caused by the increase of the armed forces was simply insane. The strength of the Hungarian People's Army, 178,266 on November 5, 1951, had risen to 190,000 by December 30, although it is interesting to see that those commanding them were declining in numbers, mostly because of the purges of officers that were taking place: the 23,800 officers on November 5 had fallen by the end of the year to 15,157 (22 generals, 93 colonels, 274 lieutenant colonels, 861 majors, 1,175 captains, 2,115 first lieutenants, 3,362 lieutenants and 7,255 subalterns).[5]

These were the conditions under which the National Defense Academy began its second academic year. Its regular strength (effective strength henceforth in brackets afterwards) on September 25, 1951, was 5 (1) generals, 623 (461) officers, 251 being students, 81 (63) subalterns, 124 (143) privates, and 228 (201) civilian employees, making a total of 1,061 (869) and 54 (24) horses.[6] (A large shortfall of officers appears because the first-year students had not arrived yet.)

A great influence on the 1951–1952 year at the Academy was exercised by a measure of the chief of General Staff, Lieutenant General István Bata, presented for study in an order of November 14, 1951.[7] This stated essentially the following:

> It must be achieved by January 1, 1952, that every higher unit, group, sub-unit, school, institute and store of the People's Army, and the entire strength of the army know the tasks in the order that devolve upon them to such an extent that the commanders and the whole personnel, in the rest of the training year, can report at any time on the specific tasks of each unit or staff under their command and those devolving on themselves personally, and on the accomplishment of these. The Comrade Minister of Defense has ordered that all officers and generals study this Order... regularly and treat it as the basic document of combat, political, operational-tactical and staff training in the 1951–1952 training year.

The other seminal document for Academy training in the new academic year was entitled "Measure on Commencement of Officer Extension Training Courses."[8] Here the chief of General Staff ordered that the nine-month Higher Commander Course for 81 persons (including 4 border guards), the one-year Regimental Commander Course for 180 students (including 7 border guards), and the Academy Classes[9] (including 10 border guards in the two-year course) should all begin on November 1, 1951. The complement report for November 15[10] mentions 5 regular (2 effective) generals, 623 (649) officers (of whom 431, including 21 from the

ÁVH—the State Security Authority—were students), 81 (58) sub-alterns, 124 (134) privates, and 228 (219) civil employees, 1,061 (1,062) altogether, and 54 (47) horses.[11]

The new students of the Higher Commander Course took an assessment examination, in which the basic knowledge of 70 percent was shown to be weak and their staff culture "very weak."[12]

The disciplinary situation was essentially unchanged: 16 officers were reprimanded, most of them for negligence in obeying orders. Only 6 received a commendation. The political situation and morale were thought to be good, the "comrade students" (as the new terminology dubbed them) showing "great industry and enthusiasm in setting about their Stalinist military studies." The Party political work was centered on giving positive form to the minister of defense's order to the Academy personnel. The comrades "unanimously reached the conclusion that they sense the strength of our army and sense also the responsibility that rests on their shoulders, to be able to utilize correctly the military equipment and the personnel placed in their charge."

On December 16, the minister of defense ordered the Academy to plan the training of Party functionaries in 1952.[13] A second group of leading Party functionaries was to be prepared between January 4 and December 15, 1952 (with a break in July), through participation in three hours of training twice a week (258 hours) and a three-day divisional exercise in September. The training goal was to teach the bases of modern combat and questions of cooperation between branches, as well as the main principles of battle under complex conditions and circumstances. This would involve the following:

- In general tactics, teaching the bases of all-branch combat and the role and operation of an infantry regiment within an infantry division, and presenting the offensive and defensive combat of an army corps;

- In staff service, presenting the work to be done on the staff of an infantry regiment or division, and the role of the commander and his staff in directing combat;
- In basic branch knowledge ("techminimum"), imparting this to a depth that would allow it to be applied creatively in general combat, and teaching inter-branch cooperation;
- In topography, imparting the use of military maps, orientation in the field, and the use of the compass and telescope;
- In army organization, presenting the organizational principles of the People's Army to army corps level;
- On the organization of foreign (imperialist) armies, teaching the organizational and battle-conduct principles of enemy armies;
- In shooting practice and knowledge of arms, presenting the handling and use of a pistol, and acquisition of at least satisfactory shooting skills.

The course was taken by 50 leading Party functionaries. A training plan in line with these training goals was submitted to the minister of defense on December 22 by Colonel Mihály Gyulai, commander of the Academy,[14] and Major Aurél Nemes, its political officer.

Meanwhile the Academy Command also worked on implementing the minister of defense's measure[15] on further military training in 1952 for the first group of Party functionaries, who had attended the course in 1951. The goals of this were as follows:
- To deepen and develop further the materials studied in 1951;
- To learn the new principles and concepts based on the new Combat Regulations and orders from the minister of defense.

The "Summary Situation Report" on the National Defense Academy in December 1951[16] gave the complement as 786, of whom 205 were officers, 454 students and 127 civilian employees (of whom 12 were supernumerary). The training of the students had

Colonel Mihály Gyulai

proceeded regularly in December. The first-year Academy Course had dealt with the defensive combat of an infantry battalion, while the second year had completed the material on divisional offensives and had begun working on divisional special combat. The Regimental Commander Course had completed the defensive combat of an infantry battalion. The Higher Commander Course had begun the whole material on the defense of an infantry regiment.

The report identified some serious disciplinary problems:

The disciplinary state of the students in relation to the teachers is not the best. They criticize a great deal and there are a lot of complaints about the disciplinary actions of the lecturers. It is particularly noticeable that students do not conduct themselves towards low-ranking lecturers in the way that subordinates should towards superiors, irrespective of their rank. The lecturers—regardless of their rank—strive to create iron discipline and to follow the practice of the Frunze Military Academy; thus it is inadmissible for students to evaluate or criticize the orders of lower-ranking lecturers, just because they carry a higher rank.

The Party functionary students in the various classes took an end-of-year examination in December. Only one or two students scored a grade of "satisfactory" (3); the remainder completed the course with grades of "good" or "excellent" (4 or 5).

The Academy Command made further efforts to raise the standards of teaching and examining. On January 21, 1952, the commander and the political officer put forward a proposal to the head of the Political Chief Section on the subject of the examination system to be undertaken at the National Defense Academy.[17] It was observed that the conduct of examinations at the Academy was not standardized, and accordingly three types of examination were proposed, in line with the examination system in Soviet higher military schools: interim, end-of-year and state examinations.

The goal of the *interim* examinations was to say whether the training goal in each training period had been achieved. Such examinations might be oral or written, or might take the form of an assessment task. The number and location of the examinations would be laid down in the training plan; the conduct and results would be organized and recorded by the Academy Command. The fact of taking an examination would be recorded in the students' examination books. This examination system would be augmented by the seminars laid down in the training plans.

The *end-of-year* examinations would conclude the first-year and second-year Academy Courses and the Regimental Commander and Higher Commander Courses. The goal of this type of examination was to establish whether the students had achieved the annual training goals of the course. They would be organized and conducted by the Academy Command, but the chairs of the examination boards of the two courses would be appointed by the chief of General Staff. The period of preparation for the examinations would be laid down in the training plan. The subjects examined would be as follows:
- combat;
- Marxism-Leninism;
- Russian language;
- military history (second-year Academy Course only);
- topography (first-year Academy Course only).

Successful examinees in the two courses would receive a "certificate" to that effect from the Academy Command.

The two-year Academy Course had to conclude with a state examination, the purpose of which was to ascertain that the students had acquired higher military training. Those successful in the examination would be classed as an "officer with higher military training" and would receive a "diploma." The Academy would issue two types of diploma. Those passing the examination with a grade of "satisfactory" or "good" would receive a plain "diploma," while those obtaining the grade of "excellent" would receive a document in which there would appear under the word "diploma" the text "with distinction." All students who had received a grade of "excellent" in all the examinations in both courses and the state examinations and in their diploma work would receive both a "diploma with distinction" and a "gold medal of the Academy" with the associated "honorary diploma." The constituents of the state examination would be as follows: diploma work (presentation and defense at regimental, divisional or corps level), combat, Marxism-Leninism and Russian language.

For admission to the state examination, students would have to have passed the end-of-year examinations of the two courses and to have defended their diploma work. The subject of the diploma work for second-year students would be chosen from a register prepared by the departments. The department concerned would appoint an adviser, to whom the student submitted a draft. Once this had been approved, the student would begin writing, turning to the adviser if difficulties arose. The manuscript would be submitted to the adviser, who would recommend corrections, omissions or additions, before it was put in its final form, at a specified time, before a three-man board (chair, "prosecution" and "defense") appointed by the department. The defense would be conducted as follows:

- The candidate would read his diploma work over 30 or 40 minutes, or pick out the salient points in it. Other students could then put questions to the candidate, criticizing or supporting what had been heard. The board members could also question him.

- The candidate would reply to the questions and "charges."

- The "defense" (the adviser) would present the positive points of the candidate and his work, after which he would request the board to assess the diploma work highly.

- The "prosecution" would recount the weaknesses and short-comings of the work.

- The candidate would exercise the right to have the last word to defend his work and respond to the "charges" or possibly con-cede his mistakes.

- After a five- or ten-minute break, the chair would announce the grade for the diploma work: "excellent," "good," "satisfactory" or "not satisfactory."

The board of the state examination was to be appointed by the chief of General Staff, and the chairs of the examination boards and sub-boards would be chosen from its members. These would be augmented by teachers from the Academy. The results of the state examination would be endorsed by the chief of General Staff. In other matters of organization, the "Full Requirement Program" devised by the commander of the Academy and endorsed by the chief of General Staff would contain the questions that embodied the subject matter of the examination. The diploma work register would also have to be submitted to the chief of General Staff for approval.

At the end of the proposal, the commander of the Academy, who had graduated from the Frunze Military Academy, and his political office emphasized the point that "the above system corre-sponds to the examination held at the Soviet higher military schools. Introducing this system will allow fair and correct adjudi-cation and encourage enhanced study, and I therefore recommend its establishment at the National Defense Academy."

The January 1952 "Summary Situation Report" on the National Defense Academy[18] stated that the training had proceeded as planned. The first-year Academy Course and the Regimental Commander Course had covered the whole material on defense of an artillery regiment. The second-year Academy Course was nearing completion of the training in special combat, while the Higher Commander Course was working on the full material for attack by an infantry regiment. Meanwhile the students had taken the general examination on basic knowledge on January 15–16; the authorities had concluded from this that "the students are weak in the field of theory. They have not learnt sufficiently… the material of relevant regulations." Although in practical work they had proved better, with the artillery class of the first-year Academy Course showing marked progress in reaching command decisions, there was a general problem for students in organizing cooperation and with orientation in the field. Behind the performance of the weaker students also lay some weaker instruction work. Most of the departments were neglecting to prepare thoroughly for methodological activities and failing to provide younger lecturers with enough assistance. Some of the lectures were not sufficiently expressive. On the other hand, most lecturers were taking part in devising and assessing the new regulations and aids.

Training in special classes also progressed. The Party functionaries received basic lectures and arms demonstrations. The State Security Authority's class studied the offensive combat of an infantry regiment. In that month its students underwent their midyear examinations, with generally good results, although absenteeism was high. Here the disciplinary situation had improved a lot. Inadmissible criticism of teachers by students had ceased. "This is attributable on the one hand to lecturers' action demanding decisive discipline, and on the other to more specific work in the commands of the year and the course."

The disciplinary situation in general was described by the Academy Command as improving, although one air force instructor, a major, had committed suicide during the month, and 12 officers (including 7 course students) had been reprimanded for lateness and for negligent performance of their duties. In the same period, 29 officers (including 26 students) had been commended for their good study results and industry.

The political situation and morale among the Academy complement were also rated as being good; indeed, the authors of the report noted that "As a result of the recent membership meeting to elect the secretary of the Party organization and the last branch meetings to elect leaderships, the disciplinary situation in the Academy has improved greatly compared with December." It was also remarked that the grades in the Regimental Commander Course examination in Marxism-Leninism and in the second-year Academy Course examination in political economy averaged about 4.

An order[19] was issued on May 14, 1952, on "Regulation of the Training System at the National Defense Academy in the 1952–1953 Year."[20] Here the minister of defense ordered that "the period of training for the new first year of the Academy branch commencing in the fall of 1952 will be three years." The following student numbers were set:

Branch	Higher Commander Course	Regimental Commander Course	New Academy First Year
Infantry	30	40	30
Artillery	15	40	15
Armored	15	20	15
Other	10	-	-
Anti-aircraft Artillery	-	20	15
Signals Troops	-	20	10
Air Force	-	20	15
Engineers	-	-	10
Engineers and Chemical Defense	-	20	-
Total	70	180	110

One important factor in raising standards was the fact that the minister ordered an entrance examination for the first-year Academy Course, unlike the two commanders' courses. This was to be held between June 25 and July 20, 1952.

The Chief Section of Personnel was given the task of organizing the selection of candidates and ensuring the prescribed numbers of students. The Academy Command was to submit by June 15 a proposal on the new organization for the institution. It emerges from the minister of defense's document[21] that the Academy had been through some tough times in May. The minister had made an inspection on May 2, during which he had received the impression that "the Command, the heads of departments and the teachers of the National Defense Academy do not sense the responsibility that they bear [for ensuring] that officers who are excellently trained, disciplined, exemplary in behavior, and living in the spirit of the regulations should emerge from the Academy. The inspection revealed that the present spirit, discipline and order of the Academy cannot guarantee that requirement." So on May 10, 1952, the minister ordered a review board to go to the Academy, to unearth the roots of the mistakes and shortcomings and end the omissions. The board concluded that although there had been improvements since the minister's visit, these could mainly be seen in outward appearances, or at most in internal order. The shortcomings found in the fields examined and the main actions to be taken were as follows:

- "The commander of the Academy should without delay... turn the Academy into an exemplary model of discipline, order and life in the spirit of the regulations, such as can, for example, serve for the whole People's Army, and ensure that the officers emerging from the Academy will make the discipline of the forces tougher still."

- The Academy had to work out its Organizational Plans by July 1.

- Harmony had to be brought to the work of the departments; they were not to tolerate liberalism in grading the students.

- The methodological preparedness of the teaching staff had to improve; no methodologically unprepared teacher might lecture; activities had to be assessed in every case; it was forbidden to hold topographical activities in the classroom. "The commander of the Academy has to end the laxity and carelessness that appears today in the style of the lectures, in the behavior at lectures, and in the notes of the lecturers."

- The development plan for cabinets (exhibition rooms for charts, maps, and so on, to aid study) had to be prepared and presented to the minister for approval by July 1.

- The commander of the Academy was to oversee the inspection plans of course and year commanders, heads of departments and senior teachers, and their implementation.

- The commander of the Academy's inspection plan had to be submitted to the minister by June 1.

- Other foreseeable events (dress parades, command exercises, and so on) were to be considered when planning the 1952–1953 academic year.

- A radical improvement had to be made in shooting practice, to produce grades of "excellent" and "good" by the end of the training year.

- An alarm plan in line with the Service Regulations had to be devised and practiced by July 1.

- An immediate list had to be made of those handling secret ("T") documents; all irregularities in the copying and destruction of "T" materials had to cease immediately; separate premises and staff had to be provided for storing and handling service books.

- The Chief Section for Organization and Mobilization, in consultation with the commander of the Academy, was to inspect the Academy's complement table, and, if need be, make a recommendation by June 10 on augmenting it or filling vacancies, or increasing the vehicle fleet.

- The Academy, in consultation with the Air Force Command, was to devise a proposal for acquiring the equipment required for training air force officers.

- The Chief Section for Operations was to examine the Academy's operation every month and report on it to the minister by the 15th of each month.

The three-man board concluded the following:
The inspection showed that the National Defense Academy meets every requirement for ending the current shortcomings as rapidly as possible. The Academy constantly enjoys the self-sacrificing support of our people, Party and People's Army. The Academy is in the happy position of being able to draw directly, through Comrade (Soviet) Advisers, on the treasure of Stalinist military science. I am sure that the Academy Command and whole secretariat will use their whole expertise, talent and strength to repair the shortcomings, making the Academy a model of discipline, order and military knowledge that thereby comes to resemble its great paragon, the splendid Frunze Military Academy.

So it was "natural" for the May monthly "Summary Situation Report"[22] to state the following almost euphorically: "The minister's inspection by the Comrade Minister of Defense on May 2 had a very positive effect on the personnel, as did the minister's inspection and subsequent assessment by the board headed by Comrade Lieutenant General Bata... It has been found in general that the behavior of the comrade students has improved since the minister's inspection, and there has been a marked improvement in the field of internal order." It was thanks to the minister's and the chief of General Staff's inspections that it had been possible to persuade the teachers "to desist from reading out their lectures like dictation, or to explain the lecture with visual material."

Like their predecessors, Colonel Mihály Gyulai, commander of the Academy, and Major Aurél Nemes, its political officer, put forward a proposal for reorganizing the National Defense Academy.[23] They called the present set-up "complicated and possessing a range of shortcomings that impede the continuity of the training and have an effect on the leadership itself." The institution, they argued, was not a "real" all-branch academy, and more an "academy group" comprising all-branch, artillery, armored, air force, anti-aircraft artillery, signals troops and engineering academies. There were two Academy years and two courses with different training goals operating within it, making altogether 25 classes of students pursuing 22 different types of training. The proposal was therefore to combine the "branch academies" into an institution "able not just to turn out middle-ranking cadres for all branches of our People's Army, but to train up leading cadres as well." The other big problem named was that all the Academy's burdens—in study, training and discipline—rested on the departments. The tension that this caused was heightened by the departments' low staff levels. The proposal, essentially, was to "find for the Academy a form of organization that would encompass an academy of all branches, and strive to implement this within three or four years... This organization will point to a time when the various courses in our Academy will have ceased and there will simply be three Academy years and a higher commanders' course operating." Those coordinating with the commander of the Academy would include the Academy political officer. There would also be two deputy commanders subordinate to him, one on the training side and one on the studies side, as can be seen in the following breakdown of the organization of the National Defense Academy for the 1955–1956 academic year.

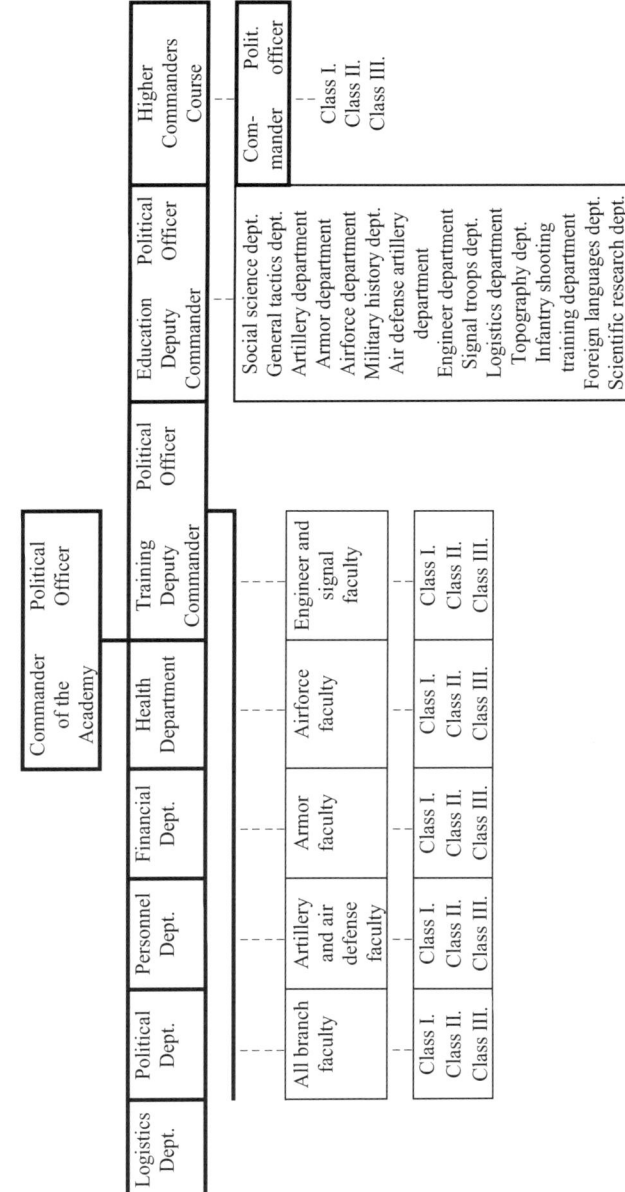

Organization of the National Defense Academy, 1955/56

A transitional form of organization was recommended for the 1952–1953 and 1954–1955 academic years. The Academy Command requested from the chief of General Staff the appointment of an organization committee to assess the proposal and report by July 1 on how to implement it.

There was also an urge to raise standards behind a measure taken by the commander of the Academy on the conduct of examinations in 1951–1952.[24] The general principle was for both Academy years and both courses to take written and oral examinations. The state examinations were to be conducted by a board appointed by the minister of defense, while the written and oral topics and questions would be devised by the General Staff of the Hungarian People's Army, in consultation with the branch commanders, or, as in the case of the social sciences, the Political Chief Section. The composition and tasks of the examination boards and sub-boards were detailed thoroughly, along with the tasks of various bodies in preparing and conducting the examinations and in decorating the classrooms, and in ensuring appearance for the examinations and appropriate dress. The memorandum exemplifies the continual efforts by the authorities at the National Defense Academy to raise the standard of Academy training and examination, under the conditions that prevailed.

The students on the Higher Commander Course completed the material on offensive combat by an army corps and turned to study of special combat methods.

The average study grades in May were as follows:

	Average in May
First Year of Academy	3.88
Second Year of Academy	3.93
Higher Commander Course	3.74
Regimental Commander Course	3.97
Aggregate for the Whole Academy	*3.88*

In June, the number of unfilled officer posts increased by 4 (to 23), while the number of vacant teaching posts was unchanged at 12.[25] One important event from the training point of view was that the students of the Higher Commander Course and the air force class of the Regimental Commander Course took part in an exercise on July 2–21 under the command of the minister. This gave students a lot of practical experience and broadened their outlook. All students, with the help of their class teachers, were preparing systematically for the end-of-year or state examinations. The departments had also assessed their individual tasks. The diploma-work defenses by the second-year Academy Course had gone as they should; although "copying also occurred on the part of students during preparation of the diploma work." The first-year Academy Course completed the subject matter of defensive combat by an infantry division. The Regimental Commander Course began to teach the offensive tasks of a division. The Higher Commander Course taught the subject of attack by an army corps and began to study special types of combat.

The average study grades in June showed no essential change since May:

	Average in June
First Year of Academy	3.95
Second Year of Academy	3.90
Higher Commander Course	3.74
Regimental Commander Course	3.86
Aggregate for the Whole Academy	*3.86*

One unexpected event in connection with the admission examinations was the fact that the armored troops sent the best-prepared candidates, which had by no means been the case earlier.

Name	Officers	Sub-alterns	Pri-vates	Civilian employees	Total
Command	4	-	-	4	8
All Branch Faculty	3	1	-	2	6
Artillery Faculty	3	1	-	2	6
Armored Faculty	3	-	-	3	6
Air Force Faculty	3	-	-	3	6
Engineers and Signals Faculty	3	-	-	3	6
Deputy Commander	3	-	-	3	6
Political Department	19	1	-	12	32
Personnel Department	5	3	-	2	10
Scientific Research Department	2	-	-	4	6
Social Science Faculty	30	-	-	2	32
General Tactical Faculty	32	-	-	9	41
Artillery Faculty	19	-	-	5	24
Armor Faculty	20	-	-	6	26
Air Force Faculty	12	-	-	4	16
Engineers Faculty	16	-	-	5	21
Signals Faculty	15	1	-	5	21
Anti-Aircraft Artillery Faculty	11	-	-	4	15
Supplies Faculty	4	-	-	1	5
Infantry Shooting Training	4	-	-	3	7
Military History and Organization	6	-	-	2	8
Topography and Mil. Geography	7	-	-	3	10
Foreign Languages and General Knowledge Faculty	1	-	-	10	11
Education Department	10	5	-	70	85
The Administrative Deputy Commander	22	68	138	53	281
Military Supplies	7	4	2	92	105
Financial Service	2	2	-	2	6
Health Service	3	2	-	5	10
Total	**269**	**88**	**140**	**319**	**816**

The political situation and morale were rated as being good by the Academy leaders. The atmosphere was typified by the "enthusiastic response to the appeal by the National Peace Council to assist the Korean people... It was emphasized that we could help Korea not only with financial assistance, but by fulfilling our tasks to the full." By July 28, 1,126 persons had contributed 80,380 forints for Korea—equivalent to 5.2 percent of the monthly payroll, although the sacrifice made by the Regimental Commander Course was thought to be insufficient; a second collection there yielded 750 forints.

The Academy Command worked hard to gain respect for the institution and raise its reputation. On September 5, 1952, for instance, the commander of the Academy, Colonel Mihály Gyulai, with a countersignature from the political officer, proposed to the minister of defense, Colonel General Mihály Farkas, that the Academy personnel should be placed in a new, higher-paid, category.[26] Gyulai also strove to raise the standard of military history teaching, writing on the matter early in September 1952 to "Comrade Minister Erik Molnár, Director of the Institute of Historical Studies,"[27] with two main requests. First, "I request, in order to raise the standard of military history teaching in our Academy, that the Institute of Historical Studies check the lectures prepared for the next training year, and assist our lecturers by remedying the deficiencies in them." The department planned 60 hours of instruction on Hungarian military history, from the Conquest to the history of the two World Wars. These lectures would be produced and sent to the Institute of Historical Studies successively between October 1952 and March 1953. The other request was for lecturers for ten hours of methodology, to be delivered to the lecturers on military history, covering the French Revolution and Napoleon, Europe from the defeat of Napoleon to the unifications of Germany and Italy, the Paris Commune, and the First and Second World Wars. He also requested lecturers for 30 hours on the subjects of Hungarian prehistory, the Conquest and raiding period, the Árpád period, the Angevin period and the peasant revolt in Transylvania, János and Mátyás Hunyadi, Dózsa's peasant revolt and its consequences, the Hungarian people's struggles against the Turks and the Habsburgs from the Battle of Mohács to the Rákóczi Wars, the 1848–1849 War of Independence, Hungary between 1849 and 1914 and Horthy's fascism. This initiative soon led to the appearance of a serviceable textbook: *Jegyzet a Honvéd Akadémia számára I-III* (Notes for the National Defense Academy) (Budapest: Hadtörténeti Intézet és Múzeum, 1954).

Instruction of the students proceeded as planned. It was reported that the end-of-year examinations had been held. The second-year Academy Course had begun the state examinations. The written tactics examination and specialist orals had been taken and the tactics orals had started on May 30. But there had been problems with the artillery students, 18 percent of whom had failed the specialist oral examination on artillery firing commands: "This demonstrates that greater attention must undoubtedly be paid to teaching artillery firing commands in the next training year." One student had failed to defend his diploma work, but it had been possible to allow all the others to enter for the state examination. The first-year students in the special classes (Party functionaries) had finished their studies of infantry division defense, had completed the seminar with good results, and had begun to study infantry division attack. The two classes had been visited on September 30 by Lieutenant General István Bata, deputy minister of defense and chief of General Staff. The lecturers were heavily burdened with preparations for the students' examinations, the month of methodology instruction, and preparation of the teaching materials for the first three months of the next academic year: "The work of the department is hindered because most of the 'Frunze Military Academy' lectures are now undergoing stylistic revision or approval. The lecturers' work is absolutely essential, so that at least one copy of all the translated 'Frunze' lectures can be found in the Academy Library."

One constant topic over the academic year was overburdening of the teachers. Nonetheless, the amount of extra work done was surprising. The 31-man Department of General Combat, for instance, was short of 12 lecturers, which placed a heavy extra burden on the others. Yet the report said that their superiors had "not spared them." One success recorded was in setting up cabinets for artillery combat, artillery firing practice, armored combat, air force combat, infantry firing practice, topography, Party history, history and political economy. It was thought necessary to stress how "all these achievements... in the last study year had been made possible by the effective, untiring work of the Comrade (Soviet) Advisers at the Academy."

There was a remark on the urgency of meeting the lecturers' housing needs in Budapest as soon as possible, so that they could devote themselves undisturbed to their important educational work.[28] Another problem mentioned was the fact that there were no quarters available at the Academy for the students on the Higher Commander Course, who had to be lodged at the Adria Hotel, whereas it was highly desirable from the study and the disciplinary points of view that all Academy students be quartered there. Perhaps even greater problems were caused by the fact that 40 teachers had no apartment in Budapest. In fact, one of them was living in a concierge's apartment and carrying out the concierge's duties (taking out tenants' refuse, opening the door at night, and so on).

Perhaps more interesting data than those found in these two "Summary Situation Reports" can be found in the document on which they were based.[29] From this the following emerges:

- Lieutenants made up the decisive majority of the first-year Academy Course (up to 80 percent), while 10 percent each were sub-lieutenants and first lieutenants. Most students came from worker (52 percent) or peasant (45 percent) families, hardly 3 percent from the intelligentsia. Almost all belonged to the 20–30 age group. 98 out of 107 had completed a one-year branch course, 8 a reserve officers' school, and 1 a sub-divisional commanders' course. As for their educational attainment, 93 had completed 6–8 years of school, 13 had completed civil (middle) school, and just 1 had graduated from high school. Over half had commanded a squad or company, or in two cases a battalion; almost a third had held a staff post and the rest had served as adjutants. Over 90 percent of them had been officers since 1949–1950, and more than 86 percent of them were Party members or probationers. In the end-of-year examinations, 13 percent received a grade of 5, 67 percent a 4, 14 percent a 3, and 6 percent a 2.

- About 80 percent of the second-year Academy Course students held the rank of first lieutenant or captain, or were sub-lieutenants or majors. Of the 62 students, 39 came from a worker, 22 from a peasant, and 1 from another background. Almost two thirds were in the 20–30 age group and over a third in the 30–40 age group. Over half had completed officers' school, and about 40 percent the sub-divisional commanders' course. Two students had completed the Regimental Commander Course and one had studied at the Stalin Academy of Military Policy in Moscow. (There is nothing specific to shed light on why the last then had to undertake this training.) Two persons had completed no kind of military schooling. In their general education, 55 percent had completed 6–8 years, 39 percent had completed middle school, and 6 percent had graduated from high school. As for their previous posts, 2 had commanded a regiment, 15 a battalion, 9 a company, 2 a squad, 1 an army corps staff, 1 a regimental staff, and 1 a battalion staff. Again about 90 percent had been officers since 1949–1950. All students but one were Party members or probationers. As for their grades, 15 received a 5 in their end-of-year examinations, 31 a 4, 15 a 3, and 1 a 2. There is no sign of a failure in the records of the state examination either, where 14 received a grade of "excellent," 34 of "good," and 12 of "satisfactory." Interestingly, the "State Examination Board" made proposals on the basis of the average results in the state examination and the end-of-year examination to its chairman, Lieutenant General István Bata, deputy minister of defense and chief of General Staff, who made the final assessment.

- The Regimental Commander Course had 188 students with a wide range of ranks: 1 lieutenant colonel, 6 majors, 20 captains, 51 first lieutenants, 97 lieutenants, and 13 sub-lieutenants. Again worker (55 percent) and peasant (40 percent) backgrounds predominated; only 5 percent came from the intelligentsia. Almost three quarters were aged 20–30, a quarter 30–40, and only 2 students 40–50. Almost exactly half had completed officers' school and half an officers' course of some kind. In educational attainment, almost 50 percent had completed 6–8 years of schooling and 44 percent the 4 years of civil school, while 6 percent had graduated from high school and 1 student (0.53 percent) had a university degree. Only 86 had commanded a unit (between squad and battalion) and 19 had held some kind of staff post; the others had served in other capacities (subaltern, adjutant, cadet, Party secretary, and so on). The vast majority had served as officers since 1949–1951, only 1 since 1948, and 3 since 1947. There were 6 non-Party members. Of the 188 students, 186 took the examination, 29 percent receiving a grade of "excellent," 44 percent of "good," and 26 percent of "satisfactory." There were 3 students with a grade of "not satisfactory" (1 percent).

- The Higher Commander Course included 1 colonel, 3 lieutenant colonels, 22 majors, 20 captains, 23 first lieutenants and 10 lieutenants. 35 percent were of worker, 40 percent of peasant, 2 percent of intellectual, and 20 percent of other background. Of 79 students, 36 were under 30, 40 aged 30–40 and 3 over 40. Previously, 51 had completed some kind of course, 22 an officers' school, and 6 no kind of military schooling. In their education, 65 had completed 6–8 years of elementary school, 28 percent had completed middle school, and 7 percent had graduated from high school. In the light of their military and civil education, it may be surprising that 2 had served as chief of section in the Defense Ministry, 3 as head of department, and 1 as deputy head of department; 8 were instructors in an officers' school, 16 battalion commanders, 12 commanders at regiment-level and 4 divisional or army corps-level commanders. The vast majority (74) had become officers in 1949–1950, only 5 being commissioned in 1948 or 1951. All were Party members. In the end-of-year examination, 11 received a grade of "excellent," 36 of "good," 31 of "satisfactory," and 1 of "not satisfactory."

The October "Summary Situation Report"[30] shows an improvement in the personnel situation since September: a shortfall of 24 officers, out of which there were 12 unfilled teaching posts. It also emerges that the post of political officer of the Academy was vacant. In training, the state examinations had ended on October 10, with generally expected results. That day had marked the end of the academic year. Methodological preparation for the teachers started on October 5, and work on the lectures was being done concurrently. In discipline, there had been 13 reprimands and 25 commendations. The political situation and morale were assessed as being good; the attention of the complement had been caught by the Nineteenth Congress of the Communist Party of the Soviet Union.

That ended the National Defense Academy's second academic year. There had been clear failures, but serious effort had been made to modernize the teaching (with diploma work, a new examination system, and so on). The groundwork for developing military higher education had been laid by introducing entrance examinations and a three-year course. It seems as if the "class war" in the Academy (to use a fashionable cult-of-personality expression) had cooled, and there was no "unmasking of imperialist agents" as there had been in the previous year. A fairly small number of 8 students out of 465 had been removed for political reasons, and with one exception, the service offenses had brought a reprimand, not arrest and sentence by a military court. The emphasis had apparently shifted onto qualitative development of Academy work, within the scope and conditions that prevailed.

Notes

1. László Csendes and Tibor Gellért, *Kronológia a Honvédség történetéből 1945–1990* [Chronology of the History of the *Honvédség*, 1945–1990] (Budapest, 1993), pp. 106–107.
2. László Csendes and Tibor Gellért, *Háborútól a forradalomig 1945–1955. Adalékok a magyar hadsereg történetéből* [From War to Revolution, 1945–1955. Contributions to Hungarian Army History] (Budapest, 1994), p. 45.
3. *Ibid.*
4. *Ibid.*, p. 46.
5. Csendes and Gellért, *Kronológia a Honvédség történetéből 1945–1990*, p. 104.
6. Honvédelmi Minisztérium, Hadtörténelmi Levéltár, Honvéd Akadémia [Ministry of Defense, Military History Archive, Military Academy] (hereafter cited as HM HL HAKAD) 0252/7/H. A.–1950 (Jan. 10, 1951).
7. *Ibid.*, 02447/Sz. t.–1951 (Oct. 21).
8. HM HL Kat. Tanint. Csfség. [Military Training Institutes Group Command] 03022/1951 (Oct. 30): the research of Dr. Antal Oroszi, colonel (ret.).
9. If the complement report of November 15, detailed later, can be taken seriously, along with the figures in the monthly "Summary Situation Report" for November, it emerges that about 160–180 students were studying in the Academy Faculties.
10. HM HL HAKAD 0252/8/H. A.–1950 (Nov. 21, 1951).
11. HM HL HAKAD 03388/H. A.–1951 (Dec. 5).
12. *Ibid.*
13. HM HL HAKAD 03599/Pk. tan. h.–1951 (Dec. 16).
14. See Biographies of Key Personalities.
15. HM HL HAKAD 03598/Pk. tan. h.–1951 (Dec. 20).
16. HM HL HAKAD 03684/H. A.–1951 (Jan. 3, 1952).
17. HM HL HAKAD 0573/Pk.–1952 (March 6).
18. HM HL HAKAD 0253/H. A.–1952 (Feb. 4).
19. This is the first National Defense Academy document that the author found in which the National Defense Force General Staff does not appear, but the situation is still transitional, as the chief of General

Staff of the Hungarian People's Army appears only in the next document. But the bureaucratic wheels turned slowly: the change of name to Hungarian People's Army had come 11 months previously, on June 1, 1951.

20. HM HL HAKAD 01158/Pk.–1952 (May 16).
21. HM HL HAKAD 01230/H. A.–1952 (May 26).
22. HM HL HAKAD 01259/H. A.–1952 (June 5).
23. HM HL HAKAD 01354/H. A.–1952 (June 12).
24. HM HL HAKAD 01239/H. A.–1952 (June 18).
25. HM HL HAKAD 01701/H. A.–1952 (Aug. 2).
26. HM HL HAKAD 01942/H. A.–1952 (Sept. 5).
27. HM HL HAKAD 01919/H. A.–1952 (Sept. 20).
28. HM HL HAKAD 02510/H. A.–1952 (Oct. 28).
29. HM HL HAKAD 02472/H. A.–1952 (Oct. 16 and 22).
30. HM HL HAKAD 02602/H. A.–1952 (Nov. 4).

Chapter VII
The National Defense Academy's Third Academic Year
(November 1, 1952–October 31, 1953)

The third academic year at the National Defense Academy was strongly affected by historic events in international politics and Hungarian public life, just as the fourth was to be.

One notable event in international politics was the decision of the Twentieth Congress of the Communist (Bolshevik) Party of the Soviet Union on October 5–14, 1952, to drop the word "Bolshevik" from its name, perhaps as a conciliatory gesture to the West. The event that stood out above all was the death of Stalin on March 5, 1953. Another one, possibly underappreciated at the time, was the Bundestag's ratification in Bonn of the treaty with the three Western powers on March 19, 1953, along with the treaty on the European Defense Community. The treaty legally ended the post-war occupation of West Germany, although the Western powers retained some important rights, and led up to Germany's accession to NATO two years later. These developments may have helped to encourage the Berlin Uprising of June 17, 1953, in East Berlin and several other East German cities, which was swiftly crushed by the Soviet occupation forces.

On the other hand, the Left was strengthened by the Italian general elections of June 1953. There was worldwide relief over the cease-fire in Korea (July 27, 1953), although war continued in Indo-China. The Western statement on August 20, 1953, about the successful Soviet hydrogen bomb test showed concern.

The succession of world events in this period, the majority of them positive, led to some easing of the Cold War tensions.

There were also a number of changes in Hungary's domestic and military policy during the 1952–1953 academic year at the

National Defense Academy. The more influential of these included completion of a stage in the development of the Hungarian People's Army, reached at the end of 1952, leaving it with the huge strength of 210,400, a big improvement in its technical equipment (T–34 tanks, assault guns, MIG–15 jet fighters, IL–10 battle planes, TU-2 bombers, and so on), 24,717 officers (25 generals, 1,492 senior officers, and 23,200 junior officers) of predominantly worker or peasant background (32 percent of worker, 21.1 percent of agricultural worker, 12.6 percent of peasant, 2.5 percent of intellectual, 16.8 percent of clerical employee, 9.9 percent of artisan or trader, and 1 percent of other origin) graduating in the Academy in 1952 represented less than 1 percent of the officer corps, and most of these had only attended courses or at most had one or two years of officers' training.[1] The army consisted of the following: two corps (five infantry divisions) and a separate infantry division; an artillery division, four artillery brigades and an anti-aircraft artillery division; three home anti-aircraft artillery divisions and a separate mixed home anti-aircraft division; two fighter and one bomber air force divisions; two engineering brigades, a pontoon-bridge battalion, a river flotilla and a chemical defense company; an armored corps (one mechanized and one armored division) and two separate heavy tank battalions.

This marked a high point in the history of the People's Army in terms of strength and organization—the maximum that peacetime Hungarian forces ever attained. Such a large army had actually been superfluous since 1950. When North Korea attacked its southern neighbor, Stalin had hoped that the West would prove to be a "paper tiger," as Mao Zedong predicted—that is, that it would not impede an armed advance of Communism. If that had been so, the next victim of Bolshevism would have been Yugoslavia. But once the West had counterattacked in Korea, Stalin had to acknowledge that the West would retaliate against an attack on Yugoslavia as well. He was not prepared to wage war on the United States, and so the attack on Yugoslavia was shelved, although this was not publicized until after Stalin's death.[2] The planned size of the armed forces was unnecessary but still maintained. Once Stalin was dead, peaceful moves towards Yugoslavia began, and with them, radical troop cuts in Hungary and other satellite countries commenced. By 1956, the army was down to a reasonable size again, and dissatisfaction among the large numbers of dismissed officers was increasing day by day, while those not dismissed lived in uncertainty. The morale of the People's Army was worsening.

On November 5, 1952, the government (known officially as the Council of Ministers) published an order reintroducing the system of individual command by abolishing the political officers' co-commander status. As the minister of defense's order of December 27 put it, "The commanding cadres in the army have now gained the knowledge requisite to orient themselves in political-type situations as well as on professional matters, and they can be simultaneously military and political commanders of the forces entrusted to them."[3] The transition to the new system had to be completed by January 20, 1953. It meant that the "co-commanding" political officers continued their tasks as the political deputies of the commanders.

Stalin's death instigated processes of an importance still unappreciated at the time, in Hungary's domestic affairs and the life of the Hungarian People's Army. The political mood was very tense in the spring of 1953. The results of pursuing a cult of personality were to be eliminated immediately, and changes made in economic and military policy. A high-ranking delegation from the Hungarian Workers' Party was summoned to Moscow, where it heard from the top Soviet leadership that Imre Nagy was to be appointed as prime minister and was to direct the introduction of radical reforms. On June 3, 1953, the Central Committee met and agreed, among other things, to reduce investments in heavy industry, raise the quantity and quality of consumer goods available, build more new housing, and separate the posts of prime minister and Party general secretary (both held by Mátyás Rákosi at the time).

The next Central Committee meeting, on June 27–28, proved even more dramatic. It dealt with what were to prove to be epoch-making issues, such as exposing errors by the Party leadership, ending shortcomings in collective leadership, eliminating the consequences of the cult of personality, decreasing Party bureaucracy, lessening state expenditure (reducing heavy industrial output and large-scale investment projects, making sizeable cuts in military spending), and improving the standard of living. Rákosi resigned as prime minister during the meeting. Colonel General Mihály Farkas was dismissed as minister of defense, on the grounds that he had been responsible personally for "developing the armed forces to a greater extent and more rapidly than necessary," while "officious dictation and an extensive cult of personality had occurred also in the army, accompanied by a line of excessive development irrespective of the load-bearing capacity of the people's economy."[4] He was being condemned, therefore, for things that had been done on orders from the Soviet Union.

In line with the Central Committee resolution, Parliament met on July 4, 1953, and elected Imre Nagy as prime minister, with Lieutenant General István Bata, hitherto chief of General Staff, as his minister of defense. A parliamentary motion on reducing the armed forces and military expenditure was passed, but specific details still had to be worked out.[5] According to a Central Committee resolution of July 22, the regular strength of the Hungarian People's Army was to be reduced by almost 25 percent, losing 6,708 officers, 8,883 subalterns, 6,699 students and trainees of officers' schools, 29,970 men, and 2,206 civilian employees, a total of 54,466.[6] It meant that the army, down to a strength of under 160,000, also had to part with relatively new officers of worker and peasant origin, which did nothing for the confidence in their livelihood of those remaining, or for the attractions of attending officers' schools or military academies. On August 20, 1953, the Presidential Council of the Hungarian People's Republic raised the National Defense Academy and the Stalin Academy of Military Policy to the rank of colleges of higher education, and on September 17 the Ferenc Rákóczi II Military Secondary School opened its doors in Esztergom, for the purpose of providing quality recruits for officer training.[7] On August 25, 1953, the government appointed the members of the Military Council of the Ministry of Defense, placing the highest-level direction of the new Hungarian People's Army in the hands of Lieutenant General István Bata (chair), Lieutenant General Sándor Nógrádi (deputy chair), Lieutenant Generals István Szabó, Károly Janza, Mihály Szalvai and Béla Székely (chief of General Staff since August 10), Ferenc Madarász, Gyula Uszta and Sándor Házi.[8] All of these were committed Bolsheviks.

So the National Defense Academy began its third academic year under truly historic international and domestic political conditions. The goal designated by its superiors for the training that began on November 1, 1953, was as follows:

> to train cadres with higher military and political qualifications for all branches and central bodies of the Hungarian People's Army—based on Stalinist military science and the experience gained by the Soviet Army in the Great Patriotic War—in a spirit of high-level discipline and loyalty to the Hungarian Workers' Party and the government. The officers shall be taught the required command and methodological expertise to train and educate troops and conduct modern combat, so that on completion of the National Defense Academy and on being posted to higher units and troops, they may live and raise their subordinates in the spirit of the regulations.[9]

Under a new three-year system of training,[10] the *First Academy Year* and the old-system *Second Academy Year* would each consist of eight teaching groups (two for all-branch and one each for artillery, anti-aircraft artillery, armored, air force, engineering and signals troops); the *Regimental Commander Course* would have ten groups (two each for all-branch, artillery and armored, in the latter one for mechanized and one for tank assault, and one each for anti-aircraft artillery, air force, engineering and signals troops); the *Higher Commander Course* was to consist of four groups (two all-branch, one artillery, and one armored). Altogether 403 students from the Hungarian People's Army and 40 from the State Security Authority (ÁVH) began their studies.[11]

The staff complement at the National Defense Academy at the beginning of the academic year, including political officers, was 253, with 338 civilian employees.[12] This meant that it was under strength by 16 officers (including 1 studies deputy to the commander of the Academy and 2 lecturers) and 17 civilian employees.[13]

Another assignment from the minister of defense was for the Academy "to ensure that students receive high-standard military and Marxist-Leninist instruction. Apart from further improving the methods of theoretical instruction, it must ensure that the Academy students gain great practical experience, so that they become good commanders and staff officers of troop and higher units."[14] This requirement prompted those responsible at the Academy to augment the training plans with subjects such as "Defense of reinforced infantry regiments (divisions, army corps) in fortified positions" and "Overrunning defensive positions with infantry regiments (divisions, army corps)." They incorporated under transport training the organization and implementation of rail transport.

Student training in November—for the First Academy Year and the Regimental Commander and Higher Commander Courses—covered the introductory lectures, while some branch-course classes began to tackle regimental and battalion defense tasks as well. The performance grades for earlier years again seem to have been considered to have improved again with hindsight, as the same refrain was repeated: "According to experience so far, the composition of the First Academy Year all-branch class is weaker in knowledge than last year's First Academy Year was."[15] On the other hand, the Regimental Commander and Higher Commander Courses were said to have a generally good composition and their students were said to work eagerly and enthusiastically. But there were grounds for concern that the arithmetical knowledge of the artillery, anti-aircraft and engineers' classes was very patchy, although they would have great need of such knowledge later in their studies. The situation was made still harder because arithmetic, physics, chemistry and technical drawing could only be taught by casual "visiting" teachers, and the timetable had to be adjusted to their other commitments.

The Academy Command was satisfied with teachers' preparations, concluding that the lectures and tasks prepared were higher in standard than before. A positive point was that the engineers were being taught according to the syllabus of the Kuybishev Academy, but the absence of an engineering Soviet adviser was regretted.

As regards the disciplinary situation, there were 17 reported offenses (10 by officers, most for infringement of internal service rules) and 53 commendations (33 to officers). The political situation and morale were rated as being good, with morale especially high among the students. For instance, Béla Berecz (a notorious Bolshevik) remarked, "There was never a month in my life when I was able to study under such conditions. I am not afraid of the difficulties; I will do all I can to attain good grades." Major Emil Bérczes[16] spoke of how "I always felt my lack of military theory in practical work with my unit. Now that I have arrived at the Academy, I would like to show my gratitude by completing the course with a grade of 'excellent.'"[17]

The complement at the National Defense Academy in December 1952 showed an improvement, with the shortages of officers and of civilian employees falling slightly from 16 to 12 and 17 to 13 respectively, but the posts of studies deputy to the commander and two lecturers remained unfilled. Also reported for the first time was a shortfall of five Party secretaries.[18] The student training continued well; their industry increased. The Party functionaries' class ended for 1952 with a final examination in December.[19] Several people soon to be prominent were among those promoted by the minister of defense on December 12, 1952 (Márton Horváth, Antal Apró, Árpád Házi, István Kossa, Ferenc Keleti, Imre Mező, János Matolcsi and Béla Biszku).[20]

The Academy Command received orders from the General Staff to "evaluate" the quality of the students admitted. The conclusions were submitted in two parts to the minister of defense during December 1952.[21] During this appraisal of the final evaluations of 220 students, 42 (19 percent) were caught out in one of the screening processes. The experience may well have astonished the Academy Command and their superiors in several ways, especially as regards the personnel work to do with registration and the blunders made by military intelligence. On the other hand, it may have been unpleasant for the decision-makers to find that a good many of those concerned had produced outstanding results and displayed great discipline.

To simplify slightly, the screening process caught people out for one of nine reasons.

1. The student had been an officer or subaltern of the Royal Hungarian National Defense Force.
2. The student had been a member of the Order of the Gallant (vitézi rend).
3. The student had been a member of the Arrow-Cross Party.
4. The student had a petit-bourgeois background and outlook.
5. The student had connections abroad.
6. The student had a criminal record.
7. The student led an immoral or drunken life.

8. The student had health problems.
9. The student was unsuccessful in his studies.

There was undoubtedly much for the authorities to think about here, for one thing on whether priority was to be given to background or to specific performance, and for another on whether it would not be a crime to appoint the future leaders of the Hungarian People's Army to high positions through such a dysfunctional system of selection. It is not known how far the conclusions reached were considered or in which cases any consideration was required. Two facts can be quoted: one is that five engineering students arrived at the Academy during the academic year and 25 were transferred away,[22] and the other is that several students were to become highly esteemed officers and generals holding high posts in the Hungarian People's Army.

A proposal was submitted on December 16, 1952, to institute a *correspondence course*[23] starting on April 1, 1953, which would give the chance of obtaining a diploma to all generals and officers of the Hungarian People's Army who

1) had already completed the Higher Commander Course of the National Defense Academy, or

2) if they had not done so, were engaged on operational/tactical training appropriate to their rank at the Ministry of Defense, the General Staff, the Political Division, the National Defense Academy, the Stalin Academy, the Branch Officers' Schools or any other central body, were organizing and heading this, and thereby were acquiring knowledge of it to such an extent as to hold out hope that the *Correspondence Course* might be successfully completed and a diploma obtained.

The course was designed to last 21 months (April 1, 1953–December 31, 1954) and to open with three sections: all-branch, artillery and armored. The syllabus would be the same as for the three-year Academy Course. The following proposals for conducting it were made:

- The students would follow the curriculum independently.

- They would meet for four-day conferences at the National Defense Academy every two months, attending seminars and group activities at which they would demonstrate the extent to which they had mastered the material. The first such conference would begin on April 1, 1953, when the students would hear a briefing about the course. This would be followed by basic lectures required for absorbing the material, then the program of requirements for the course, which would assist them in covering the material independently. The other four-day conferences would come in May and June, followed by November and January. It was recommended that they should be held in May and July 1954 for the third semester.

- Every six months—in September 1953, and March and September 1954—there would be a final six-day conference, at which the students would take written and oral examinations in their term's work.

- It was recommended that the subject matter for the diploma work would be advertised on July 1, 1953, so that the subjects chosen could be elaborated and submitted to the National Defense Academy for adjudication by July 1, 1954. The defense of the diploma submissions would take place in August and September 1954.

- The last three months of the course (October–December 1954) would be available to the students for preparing for their state final examinations and for taking them.

The correspondence course's students would have to cover the subjects of defense and attack and special combat methods, at division and corps level. Apart from learning the theory of these, they would have to compose a study of their own and defend it before the appropriate Academy department.

The Academy Command submitted a report to the minister of defense on December 18, 1952, on the state and Party functionaries who had completed the first year of their studies on December 12.[24] The course was completed by 52 people. The absenteeism was greatest among the higher functionaries.

An important initiative for the development of the National Defense Academy came on December 26, 1952, when Lieutenant General Károly Janza, deputy minister of defense and head of the Logistics Service, called on the Academy Command to draw up a detailed plan for a Logistics Faculty, to begin operating in the 1953–1954 training year.[25] This would provide a three-year course, with an expected 25–26 students for high-level studies in military supplies, which would qualify them for logistics staff posts at division, corps and army level. The proposal was to be submitted by January 25, 1953, to Janza, who would put it before the Defense Force Council.

An initiative was taken on December 29 to further the theoretical skills of the top command in the Hungarian People's Army.[26] Lieutenant General István Bata, chief of General Staff, acting on instructions from General of the Army Mihály Farkas, decided on the establishment of Special Group 3, to provide higher military training to Lieutenant Generals István Szabó and Géza Révész, Major Generals László Piros and Pál Ilku, and Colonels Károly Kutika and Ferenc Hídvégi. This was to run for 32 weeks between January 1 and September 1, 1953 (with two weeks in reserve), for four hours twice each week, making a total of 256 hours (16 in reserve). The stated objective was as follows: "To allow the students to learn, on the level of an infantry division, all-branch combat, the role and place of the special branches and technical equipment in the order of battle or support of an infantry division, and the use of these in organizing and commanding the all-branch combat of an infantry division. To convey to the students the offensive and defensive combat of an infantry army corps, through the medium of lectures."

Colonel Mihály Gyulai, commander of the Academy, submitted the proposal for a correspondence course to the minister of defense on January 20, 1953.[27] This largely agreed with the ideas of the acting deputy commander for studies described earlier. The commander of the Academy emphasized the following:

> Institution of the Correspondence Course has become necessary because the vast majority of our generals and officers in higher command posts lack a diploma, while on the other hand, annually increasing numbers of officers pass out from the National Defense Academy with diplomas for various branches, which will lead to [a situation] where there are ever more graduates, in various higher commands and central bodies, whose commander will not have such a diploma... The minister of defense of the People's Army, the head of the Political Chief Section, and the chief of General Staff of the National Defense Force will likewise receive a diploma from the National Defense Academy within the framework of the Correspondence Faculty, with the difference that they will write their diploma work on a subject of their own choosing and defend it before a committee selected by the Comrade Minister of Defense.

However, it was recommended that the course should last 2 years instead of 21 months. On participation, the commander of the Academy made a gesture towards the officers serving with the troops by opening the course to higher commanders who had not completed the Higher Commander Course but were dealing with operational and combat training. The first conference was planned to last 15 days, and it was proposed that the diploma work should be defended at the beginning of the three-month conference. The document went further than the original proposal in providing students with time for study, proposing a ministerial order granting the correspondence students two whole days a week for study, and apart from that, at least four hours three times a week. The subjects of study on the correspondence course would be Marxism-Leninism, operational and tactical training, the study of foreign armies, the history of the art of war, and military geography. State examinations would have to be taken in Marxism-Leninism (oral) and in general combat and the Russian language (written and oral). It would be necessary to establish a Correspondence Faculty with a faculty commander and two each of the best lecturers in all-branch studies, artillery and armored warfare. The vacancies created in the other faculties would have to be filled.

The shortcomings of training in the Hungarian People's Army had been criticized strongly by the minister of defense in 1952. The commander of the National Defense Academy judged that some aspects of the criticism concerned problems that impinged on the work of the Academy, and took measures at the beginning of 1953 to overcome the problems.[28]

The situation report for January 1953[29] mentioned a shortage of only six Party secretaries, three political deputies, two lecturers, one dentist and ten civilian employees. Student training had proceeded as planned. Students' study averages showed some variations compared with December.

Faculty	December	January
All-branch	3.86	3.84
Artillery	3.87	3.83
Armored	3.5	3.7
Air Force	3.7	3.8
Engineering and Signals Troops	3.62	3.93

The political situation and morale were rated as being good. The government resolution and Ministry of Defense order on individual leadership had been presented at a meeting of officers, in the presence of Lieutenant General Sándor Nógrádi, deputy minister of defense and head of the Political Chief Section, and his deputy,

Major General Jenő Hazai. "It emerged from the contributions that many had not understood the importance of individual leadership." However, "the comrade lecturers had also recognized... the importance of the order and [of] the task of raising really excellent individual commanders." On the matter of discipline, it was reported that 49 offenses (of similar underlying natures to those before) had been committed and 24 commendations had been issued.

In February, the shortage in strength rose to 14 officers and 13 civilians.[30] In training, one positive point mentioned was the rising standard of faculty extension training courses, although the crowded work schedules meant that there was little time for this. An attempt to overcome that problem would be made by introducing "study half-days." In the same month, high school correspondence teaching was introduced for those serving at the Academy. After some initial difficulties, this became popular: "The comrades are studying with great effort." The February study averages showed a slight improvement:

Faculty	February
All-branch	3.84
Artillery	3.83
Armored	3.7
Air Force	3.93
Engineering and Signals Troops	4.005

The Communists of the All-branch Faculty chose the following slogan for April 4, the anniversary of Hungary's liberation in 1945: "Forward to Grade 4 Study Results." However, this great urge to study was not characteristic of every student. The head of studies for the No. 3 Special Course, Lieutenant Colonel György Gráber, reported directly to the chief of General Staff that Major General László Piros had not attended a single class and Colonel Károly Kutika only two.[31] He therefore asked for clarification from Lieutenant General István Bata on "whether the aforementioned comrades are still to be declared members of the group and whether the requisite secret materials are still to be forwarded to them."

The chief of staff made the move on May 22, 1953, to establish what was officially named the "Correspondence Faculty alongside the National Defense Academy of the Hungarian People's Army."[32] This basically agreed with the recommendation from the commander of the Academy,[33] except that it specified the range of participants as volunteers in staff command or equivalent posts, who had to return their application forms through official channels to the chief of General Staff by July 1, 1953. To the subjects proposed by the Academy were added matters of branch (artillery, air force, armored, signals troops, telecommunications and logistics) combat, branch tactical-technical knowledge and the Russian language. Branch knowledge was also added to the subjects in which a state examination had to be taken. In a departure from the original proposal, the order from the minister of defense allowed the correspondence students one day and two free evenings for study, apart from the ordinary free days.

The political situation and morale in May were rated as being good. Only one person had not voted for the People's Front candidate in the general elections; this was taken to mean that there was 99.025 percent support for the "system" at the Academy. All the staff attending the high school course had been given time off to prepare for the examinations. On the material side, the main trouble had been that the students could only complete part of the tank-driving practice because the tanks were in poor mechanical condition. Even periodic "loans" from Tata did not help much, as the equipment from there was "very worn out" as well.

"With a few exceptions, the relations of the Army Branch Commands with the National Defense Academy cannot be called good. In general, the Branch Commands only inquire into the National Defense Academy's affairs at the end of the academic year. Such interest... is generally confined to knowing how many officers the branch will receive." The commander of the Academy recommended that "every quarter the Army Branch Commands should review the progress of teaching at the Academy and report their observations to the chief of General Staff."[34]

On June 25, the Academy Command made a submission to its superiors on preparations for the new academic year.[35] The introduction noted that much trouble had been caused during the current training year by the lack of a clear statement from its superiors on the system of goals and tasks. There were too many improvisations and changes in mid-stream. "If the National Defense Academy is to begin the new academic year on November 1, it has to be in possession by that time of a full, detailed plan for the whole academic year, for without it the academic year can neither be started nor carried out. For the plans to be completed in good time, all the data covering the training prospects for the whole People's Army in the coming year must be available to us, at least by July." The practice at the Frunze Academy was given as an example. Every student there received at the beginning of the academic year a "calendar" for the whole of it. "The National Defense Academy, in the forthcoming training year, must begin the year in a similar fashion if qualitative change is to occur in the knowledge and teaching-methodology work of the lecturers, and in the knowledge and iron discipline of the students." The Academy requested to this end that it receive the following data during July:

- Separate training instructions for each all-branch and branch student in each year and on each course, including the training goals and tasks. This should come not just from the General Staff, but also through the same from the Branch Commands.

- A precise plan of the employment of the students and teachers, indicating which months they would be absent and for how many hours.

- A plan of what special types of combat training should be given in the People's Army, to allow the requisite hours to be entered correctly in the training plan.

- It was necessary to settle the relationship between the branch commanders and the branch heads of faculty, to end the unhealthy practice whereby branch commanders were able to dispose directly over the branch heads of faculty and their teachers. "That is also how it came about that the commander of the armored forces... called the senior ballistic teacher of the Armored Faculty to account at the Esztergom firing range, for the inadequate firing results of six students, in such a way that the senior teacher lay senseless for an hour and proved useless for the rest of the placement due to his state of health, whereas previously he had performed his tasks with exemplary effort."

- It was necessary to decide at the start of the academic year roughly what percentages of the students on the one-year course would go to what type of posting, so that the teaching could be more focused and the Academy thereby more efficient.

Work began on a draft for the military history textbook, in conjunction with the Archives, and on the end-of-year examinations for instructors pursuing high school studies. The political situation and morale were affected badly because the General Staff called on more teachers than planned for the staff commanders' exercise, at a critical time when the students were preparing to take their examinations and to defend their diploma work, and the high school and end-of-year examinations were being held.

The effects of the end-of-year spurt were apparent, as the July study grades in all the faculties improved:

Faculty	July
All-branch	4.03
Artillery	Not available
Armored	4.00
Air Force	4.09
Engineering and Signals Troops	4.21

The students of the Second Academy Year continued to defend their diploma work successfully, 18 receiving a grade of "excellent," 25 of "good" and 6 of "satisfactory." However, one student had his paper handed back for revision because "most of the work was copied." The drawing assignments in the Engineering Department were of a high quality as well: 5 "excellent," 5 "good" and 1 "satisfactory." There was a new disciplinary development when a succession of thefts was reported from the students' quarters. The cleaning staff were replaced or reassigned. There were 63 offenses and 58 commendations in July, but it was stated in the report that "some commanders punish halfheartedly and try to replace punishment with individual appeals." The political situation and morale were judged to have been good, although "we also detected a statement antagonistic to the government program and the speech of Comrade Rákosi. The Political Department is dealing with the investigation of this and those involved." The Korean armistice agreement, on the other hand, was assessed correctly and no pacifist statements occurred."

On August 5, 1953, the command of the National Defense Academy submitted a 17-page proposal[36] to the chief of General Staff,[37] to do with the Academy's 1953–1954 and 1954–1955 academic years. The reason behind the proposal, according to Colonel Mihály Gyulai, was that "such vast assignments have fallen to the National Defense Academy during the development of the People's Army that they call for an examination from all sides of the desires and demands of the National Defense Academy and implementation of these where justified. In line with the altered training requirements put before the National Defense Academy, it is necessary to change and to some extent reorganize the training system, objectives and training tasks of the National Defense Academy." These were summarized in eight chapters.

Chapter 1. The Objectives and Tasks of Training
The objectives were as follows:
- To supply to all branches and central institutions officers of the Hungarian People's Army with higher attainments;
- To instill Stalinist military science;
- To discover and examine the progressive traditions of Hungarian military history;
- To compile a Hungarian military education textbook; to pursue and develop further the field of military education;
- To collect and utilize in teaching the experiences of World War II;
- To create a uniform, correct, Hungarian-style military idiom by compiling a dictionary of military terminology.

The tasks were as follows:
a) For the three-year Academy Course:
- To teach and practice with students the organization and command of up-to-date all-branch combat;
- To develop students' theoretical and methodological knowledge of the training of troops;
- To make students capable of utilizing the branches effectively, organizing cooperation among them, and commanding them;
- To enable students to fill troop and higher command posts and staff posts.

b) For the two-year Higher Academy Course:
- To perfect their knowledge as generals and officers to such an extent that, having completed the course, they could "raise their knowledge to an Academy level" and become officers with higher qualifications.

Chapter 2: The Process and Reorganization of Teaching
The commander proposed eliminating the Higher Commander and Regimental Commander Courses at the end of the 1952–1953 academic year. The basic type of training would then be the three-year

Academy Course, but it was thought necessary to retain the two-year Higher Academy and Correspondence Courses temporarily as well.

The Academy authorities rightly saw that it was not worth having classes of over 15 (they insisted on this) and that classes of 5–6 were "uneconomic," although they did not think that it was possible to merge the small branch groups for educational reasons.

Since the study requirements had increased, the Academy Command thought that it was necessary to have an entrance examination for both types of Academy Course. On the other hand, it is unclear why they did not propose the same for the Correspondence Course, which involved learning the same material over a shorter period and while working. Colonel Gyulai undertook that by November 1, 1953, the Academy would devise and submit the material for the entrance examination, so that it could be issued to the applicants in December.

The proposal then turned to the teaching staff.

a) In the 1953–1954 academic year, the tasks for the National Defense Academy would place extra demands on the teaching staff, which would have to be increased for the next two academic years. Although there would be 24 classes in the 1953–1954 academic year, the Correspondence Faculty had to be created, with a teaching staff of 17 of the best instructors from the other faculties. So some deterioration in quality was inescapable, but the shortage could be made up from the present second-year students. The logistics class that was starting meant that the Logistics Department would have to expand from four to eight teaching staff, on which an undertaking had been received from the Hungarian People's Army's head of Logistics. An increase from four to seven in the teaching staff of the Military History Department would also be required if the teaching and the new textbook-writing tasks were to be completed. All these factors meant that the teaching staff would have to increase by 28 in the 1953–1954 academic year.

b) In the 1954–1955 academic year, the Academy would reach its greatest number of students, and the number of teachers would have to increase by a further 35. To this would have to be added 5 civilian teachers for the high school subjects. Part of the teacher shortage, as mentioned earlier, could be met from those graduating in 1952–1953.

Chapter 3: Settling Relations between Branch Commands and the Command of the National Defense Academy
The National Defense Academy was directly subordinate to the chief of General Staff of the Hungarian People's Army, and therefore the Army Branch Commands could not give direct instructions or orders to the Academy.

However, it was necessary for the Army Branch Commands to give assistance in setting up the specialist cabinets and providing teaching aids. This was especially necessary for the Artillery, Air Force and Logistics Faculties.

The branch faculties, on the other hand, had to give assistance to the Army Branch Commands, especially in drawing up plans.

Chapter 4: Demands Made on Teaching Staff and Students of the National Defense Academy during the Year
a) Demands on students: To ensure that the education proceeded as planned, it was proposed that students of the three-year Academy Course should not be called upon for any participation in exercises outside the Academy, such as staff command exercises, war games, or other central events. However, to prevent the students from being completely divorced from their units, Colonel Gyulai proposed that second-year students who had finished their end-of-year examinations, along with those studying at Soviet academies and the second-year students returning home for troop exercises, should do one month's troop service under the command of the General Staff. This would be incorporated into the

Academy's training plan, so that it would not cause any problems. However, the Academy would certainly urge that the students in their activity should be classed as troop commanders, because only experience of that kind would be of use in this education. The Academy Command also supported the participation of the two-year Higher Academy Course in staff command exercises outside the Academy, with the prior requirement that the dates of these should be received by August 20, 1953, and should appear in the training plan.

b) Demands on teachers: Colonel Gyulai considered it essential for the teachers also to maintain close ties with their troops, but in a way not detrimental to the teaching. He proposed that, on occasion, one or two persons from each faculty would be drawn into preparing, devising or carrying out army war games and staff command exercises outside the Academy. Again he stipulated that the dates should be available to the Academy by August 20.

Chapter 5: The Dates of Commencement
and Conclusion of Teaching
The academic year for the three-year Academy Branch and the two-year Higher Academy Course would start on November 1. In both cases, the first year would end on September 30 and students would return to the Academy after a month's leave on November 1. Second-year students on the three-year Academy Course would complete their year on August 31 and return on November 1 after the September troop exercise and a month's leave. Third-year students and the second-year students on the Higher Academy Course would finish their year on October 31 after the state examinations, take their leave in November, and take up new postings on December 1. The Correspondence Course would begin on January 1, 1954.

Chapter 6: Ensuring the Teacher Requirements
of the National Defense Academy in the 1954–1955 Academic Year
Two alternatives were suggested:

Version A. If the cadre situation allowed, it would be useful to meet the teacher requirements of the Academy from out of students completing their studies in 1952–1953, so that one year would be available to them for preparation.

Version B. If Version A were not possible, those who could be considered for teaching posts should be registered and join the Academy on August 1, 1954, giving them at least three months' preparation.

Chapter 7: Reinforcement of the Military History Faculty
After a ministerial inspection on April 14–15, 1953, the minister of defense ordered the Academy to pay much greater attention to teaching military history. This meant that a textbook would have to be written and a collection of outlines made, for which an extra three teachers would be needed in 1953–1954 and a further two in 1954–1955.

Chapter 8: Reorganization of the Research Department
at the National Defense Academy
The other major tasks assigned after the ministerial inspection were the writing of a textbook of military education, and the reorganizing and streamlining of the work of the Academy Scientific Council. The Research Department was to be reorganized to cover all Academy research work, under the supervision of the commander of the Academy, something that it was not yet capable of doing, as it was headed by a medical doctor and staffed by interpreters who simply did translation work. So a five-person research department had to be created to compile a textbook on military education and research and gather all the literature on military matters to be found in Hungary.

Major General Endre Matékovits

Major General Endre Matékovits, head of the Chief Section for Operations of the General Staff, gave instructions on the final examinations for 1952–1953 in a measure dated August 21.[38] There were to be written and then oral examinations for the first and second years of the Academy and for the Regimental and Higher Commander Courses. For the written examinations, there would be one or two days' preparation for the combat; the task would generally have to be accomplished in eight hours, so that in general there would be five (or four) hours available for decision-making at the appropriate level and three (or four) hours to devise preliminary or specialist measures and arrive at the combat orders. There would be a three-hour written examination in the Russian language in each class. Three to five days' preparation before the oral examinations had to be given for Marxism-Leninism, combat and branch studies, and one or two days for other subjects.

A memorandum on the shortcomings exposed by the minister of defense's inspection held on April 14–15, 1953, described in detail the measures that had been taken.[39] The commander of the Academy had issued an order on July 9 assigning specific tasks to the various parts of the Academy; further specific tasks of the moment had been assigned continually in the plan of action for each month, for each to incorporate into its monthly plan.

Meanwhile requirements had been submitted to the Academy's superiors, for instance, that 80 percent of the lectures in the General Tactics Faculty should be translations of materials from the Voroshilov and Frunze Academies; similar solutions were needed in the branch faculties as well.

On September 1, the Academy submitted to the chief of General Staff its proposals for the state examinations and the examination questions.[40] The written state examinations would be held on October 1–3, 1953, the specialist examinations for branch students on October 6–8, and the orals on October 6–19. The overall examining board would be chaired by the chief of General Staff and the examining boards by people of his choosing, the members being Academy teachers.

On September 8, 1953, the commander of the Academy submitted the guidelines for the 1953–1954 academic year for approval by the chief of General Staff.[41] These essentially repeated the objectives and tasks detailed in the proposal of August 5.[42] Colonel Gyulai also tried to define some spheres of authority. He laid down the stipulation that the teaching and training of the students and teachers would be planned and organized by the commander of the Academy personally, by his political and studies deputies, through the Studies Department. He coordinated educational, methodological and training work in the departments and faculties, bearing in mind the specific features of the branches, and supervised the course and efficiency of the activity there. He organized cooperation among the departments and faculties, and made a monthly assessment of the work there and in the other organizations, identifying the achievements and shortcomings and the ways of overcoming the latter. The departments, under their department heads, carried out the instruction and education of the students based on the approved training plans; they led them from the professional and methodological point of view, oversaw the teachers' work, and conducted work in military science. The faculty commanders, according to their branches, led the political, moral and disciplinary training of the students, organized the students' independent course of study, ensured the material bases for the training, and, under the direction of the Political Department, organized and led the Party political mass work of the students.

On September 25, 1953, the National Defense Academy received the ministerial order laying down the number of students in the 1953–1954 academic year.[43] There were to be 160 students in each year of the three-year basic Academy Course.[44] The document still mentioned a one-year Higher Commander Course for 40 people and a 40-student Correspondence Course. In line with this there was issued the autumn reorganization measure for 1953,[45] stating that the reorganization had to be accomplished by November 15, teaching posts filled by November 1, and other officers' posts filled by November 5. The new complement meant that 4 generals, 254 officers, 75 subalterns, 100 privates, 520 students and 176 civilian employees—a total of 1,128 places—and 6 horses were available to the National Defense Academy.

Major General Béla Székely, chief of General Staff, issued a Ministry of Defense order on November 3, 1953, assessing the 1952–1953 training year and assigning the tasks for 1953–1954.[46] The National Defense Academy had to absorb this by November 29, evaluating the year and assigning the tasks throughout its own structure. The order had to be treated as the basic document for the 1953–1954 year on combat, political, operational-tactical and staff training.

On November 6, 1953, the Guidelines for the National Defense Academy approved by Lieutenant General István Bata, minister of defense, were sent.[47] This was essentially a synthesis of the submissions by the Academy described in this chapter and endorsements of them.[48] It stated the following: "The present situation of our People's Army and the standard attained in training make it possible and necessary for the National Defense Academy to adopt fully the objective stated on its foundation, of ensuring cadres with higher military and political qualifications for all branches of the Hungarian People's Army and its central bodies."

Its *objectives* on that basis were therefore as follows:

- To provide officers with higher military training for all branches, based on Stalinist military science and experience gained by the Soviet Army in the Great Patriotic War;

- To train officers in a spirit of high-level discipline and loyalty to the Hungarian Workers' Party and the government;

- To teach the officers the required command and methodological expertise to train and educate troops and conduct modern combat.

The *tasks* of the Academy were as follows:

- To impart to the students the theoretical bases of Marxism-Leninism, the history of the Communist Party of the Soviet Union and Hungarian Workers' Party, the history of combat of the Hungarian people, and the bases for building socialism in Hungary;

- To teach the underlying principles of planning, organizing and commanding modern all-branch combat, the use of weaponry in modern warfare, and organization and accomplishment of cooperation among the branches;

- To ensure for staff members and commanders of troops and higher units the necessary field experience for them to perform their duties.

To meet those objectives and tasks, the *structure* of the National Defense Academy was to have the following groups:

a) The basic faculty of the Academy was the three-year group, whose training objective and tasks were as follows:

- To develop the theoretical and methodological knowledge of students to be able to perform command and staff duties;

- To impart to students the foundations necessary for high-quality military erudition;
- To raise the level of the students' Marxist-Leninist knowledge to the standard desired of higher officers.

The Guidelines plainly stated the following: "Those who can be admitted to the three-year basic Academy Course are officers who have done at least two or three years' service in the field, at a central organization or at institutes of military education. They have completed infantry or other branch schools and worked in company command or higher posts, receiving a grade of 'satisfactory.' Age limit: 32 years. The students selected for entry must have passed the admission examination to the Academy."

b) The objective of the Academy extension training course, according to the Guidelines, was to perfect the knowledge of generals, commanders of higher units, and other officers in higher posts, so that they became capable of directing and commanding the troops in line with modern requirements. This training task was defined as follows:
- To perfect in each branch the theoretical knowledge and practical abilities of the students in a division/corps context, and in organizing and commanding modern all-branch warfare;
- To acquaint the students with the bases of offensive and defensive operations in an all-branch army;
- To elevate the Marxist-Leninist knowledge of the students to a degree necessary for a higher commander.
The course was ordered only for all-branch, artillery and armored training, not for the eight branches recommended in August. The following basic knowledge was prescribed as teaching material:
- In infantry classes: defense and attack of infantry regiments, brigades, divisions and corps, under normal and special conditions;

the bases of offensive and defensive operations in an all-branch army; offensive and defensive combat in a mechanized division or corps, or a tank division.

- In artillery classes: the same as above, but attention had to focus mainly on artillery and anti-aircraft artillery support for higher units.

- In armored classes: defense and attack by a mechanized regiment, division or corps, a tank regiment or division, or an artillery division, and the bases of defensive and offensive operations in an all-branch army.

The course was open to divisional commanders and deputies, and generals and officers in equivalent posts, who needed to perfect their knowledge. Regimental commanders could also be taken in exceptional cases—if worthy of higher posts.

c) Several important documents on the Correspondence Course had appeared during the year,[49] and so the Guidelines dealt with it very briefly. The training objective was to provide an Academy qualification for officers and generals whose posts meant that they could not leave their work in the field or at a central body, but who met the entrance requirements. The training task was to ensure that students mastered the three-year basic Academy Course, while receiving all-branch and branch training. On completion of their studies, the correspondence students had to take the state examination. The most forward-looking development was the fact that the period of correspondence training was set at three years, longer than the previous Academy recommendation. The main methods of teaching were to be individual tuition and conferences.

As for the scientific work at the National Defense Academy, it was stated that the Scientific Council would include the political and the general studies deputies of the commander of the Academy. There was no specific mention of faculty commanders or persons appointed by the chief of General Staff. On the other hand, the named composition of the Scientific Council was to be put before the chief of General Staff for his approval before each academic year.

Finally, the Guidelines stated that "the National Defense Academy is directly subordinate to the chief of General Staff of the Hungarian People's Army," while the Army Branch Commanders were not.

On November 25, 1953, the commander of the National Defense Academy put his final summary report for the 1952–1953 academic year before his superiors.[50] The 403 students who had arrived from the Hungarian People's Army in March 1953 had been augmented with 5 engineering students, but 25 persons altogether had been dismissed or transferred before the end of the academic year. The social backgrounds of the 383 members of the student body that had stabilized included 198 former workers (51.6 percent), 128 poor peasants (33.4 percent), 8 kulaks (2 percent), 2 intelligentsia (0.8 percent), 28 petit-bourgeois (7.2 percent) and 19 other (5 percent). Of these, 360 were members of the Communist Party (94 percent), 6 were candidates for membership (1.6 percent), and 17 were outside the Party (4.4 percent).

The 1952–1953 final examinations were taken by 416 students, with an average grade of 4.02.[51] The weakest results came from the Regimental Commander Course (3.9) and the strongest from the Higher Commander Course (4.19). Looking at the results in detail, 80 students received a grade of "excellent" (19.2 percent), 200 a grade of "good" (48.2 percent), 127 a grade of "satisfactory" (30.5 percent), and 9 failed (2.1 percent). These proportions improved on those of 1951–1952 by a few percentage points. Of the 103 students who submitted diploma work, 34 defended it with a grade of "excellent," 46 with a grade of "good" and 22 with a grade of "satisfactory," while 1 student failed. All students of the second Academy year took the state examination, 14 of them

National Defense Academy   Appendix No. 2
1952/53 year   OH23/H.A.1953
A c c o u n t   Confidential!
of the 1951/52 and 1952/53 end-year examination results   Copy No. 3

Ser. no.	Year / course	Roll		Average result		Excellent			
						1951/52		1952/53	
		1951/52	1952/53	1951/52	1952/53	No.	%	No.	%
1.	1st-year Academy	108	138	3.98	3.98	14	12.9 %	19	13.8 %
2.	2nd-year Academy	62	103	4.02	4.04	15	24.2 %	23	22.3 %
3.	Regimental Commanders' Course	186	130	4.01	3.90	53	28.5 %	23	17.7 %
4.	Higher Commanders' Course	79	45	3.81	4.19	12	15.2 %	15	33.3 %
5.	National Defense Academy	435	416	3.95	4.02	94	21.6 %	80	19.2 %

Ser. no.	Year / course	Good				Satisfactory				Not satisfactory			
		1951/52		1952/53		1951/52		1952/53		1951/52		1952/53	
		No.	%	No.	%	No.	%	No.	%	No	%	No	%
1.	1st-year Ac.	71	65.7 %	72	52.2 %	17	15.7 %	41	29.7 %	6	5.6 %	6	4.3 %
2.	2nd-year Ac.	31	50 %	53	51.5 %	15	24.2 %	27	26.2 %	1	1.6 %	-	-
3.	Reg. Com. Course	81	43.5 %	54	41.5 %	49	26.4 %	50	38.5 %	3	1.6 %	3	2.3 %
4.	Higher Com. Course	37	46.8 %	21	46.7 %	29	36.7 %	9	20.0 %	1	1.3 %	-	-
5.	National Defense Academy	220	50.6 %	200	48.1 %	110	25.3 %	127	30.5 %	11	2.5 %	9	2.2 %

Prepared in 3 copies.   Prepared by: Capt. Pál Naményi.
1st copy: Minister of Defense   Typed by: Rozália Máté, civilian employee
2nd copy: Chief of General Staff   to dictation.   [Stamp] OWNED BY
3rd copy: Folder   November 20, 1953   MILITARY HISTORY ARCHIVES
                                         Duplicator diary serial no.: 2674.

Year 1952/53

01723./H.A.1953.
Confidential!
Copy no. 3

A c c o u n t

of the results of the state examinations

Ser. no.	Branch	Roll		Excellent		Good		Satisfactory		Not satisfactory		Average	Notes
		No.	%	No.	%	No.	%	No.	%	No.	%		
1.	All-branch	35	100%	8	22.9%	14	40%	13	37.1%	-	-	4.02	
2.	Artillery	14	100%	2	11.4%	6	42.8%	6	42.8%	-	-	4.19	
3.	Anti-aircraft	9	100%	3	33.3%	3	22.2%	3	33.3%	1	11.1%	3.97	
4.	Armored	14	100%	2	14.4%	6	42.8%	6	42.8%	-	-	4.07	
5.	Air-force	8	100%	1	12.5%	4	50%	3	37.5%	-	-	3.71	
6.	Engineers	10	100%	1	10%	4	40%	5	50%	-	-	4.00	
7.	Signals	12	100%	5	41.7%	6	50%	1	8.3%	-	-	4.56	
8.	Academy II	102	100%	22	21.6%	42	41.2%	37	36.3%	1	0.9%	4.07	
9.	NDA 1951/2	61	100%	14	26%	34	52.6%	12	19.8%	1	1.6%	4.04	

Prepared in 3 copies.
1st copy: Minister of Defense
2nd copy: Chief of General Staff
3rd copy: Folder
Prepared by: Capt. Pál Naményi.
Typed by: Rozália Máté, civilian employee
to dictation.
November 20, 1953
Duplicator diary serial no.: 2675.

receiving a grade of "excellent" (26 percent), 34 of "good" (52.6 percent), and 12 of "satisfactory" (19.8 percent). The Academy average was 4.04.

That brought to an end the third academic year at the National Defense Academy. It could be seen that the commander of the Academy, Colonel Mihály Gyulai, and his immediate staff had made huge efforts, and had had successes in many areas of modernizing and raising the standard of higher military education. A succession of measures had been taken to do with modernization of the teaching materials, examination of the students, and the educational structure, which provided foundations for subsequent progress. As a result, the Academy was given a good start in preparing and commencing the following academic year.

Notes

1. László Csendes and Tibor Gellért, *Kronológia a Honvédség történetéből 1945–1990* [Chronology of the History of the *Honvédség*, 1945–1990] (Budapest, 1993), p. 118.

2. For the development of the Yugoslav–Soviet relationship from Tito as Stalin's first disciple and Yugoslavia as a brave ally to consideration of Tito by the Bolsheviks as the main enemy, which therefore should be destroyed, see the account of one of Tito's closest associates: Milovan Djilas, *Fall of the New Class* (New York, 1998), pp. 27 ff.

3. László Csendes and Tibor Gellért, *Háborútól a forradalomig 1945–1955. Adalékok a magyar hadsereg történetéből* [From War to Revolution, 1945–1955. Contributions to Hungarian Army History] (Budapest, 1994), p. 74.

4. *Ibid.*, p. 70.

5. Csendes and Gellért, *Kronológia a Honvédség történetéből 1945–1990*, p. 132.

6. Csendes and Gellért, *Háborútól a forradalomig 1945–1955*, p. 76.

7. Csendes and Gellért, *Kronológia a Honvédség történetéből 1945–1990*, pp. 133–134.

8. *Ibid.*, p. 133.

9. Honvédelmi Minisztérium, Hadtörténelmi Levéltár, Honvéd Akadémia [Ministry of Defense, Military History Archive, Military Academy] (hereafter cited as HM HL HAKAD) 01723/H. A.–1953. (Nov. 25).

10. Dr. Lajos Móricz, "A magasabb tisztképzés magyarországi történetének elvi, szervezeti és vezetési kérdései" [Theoretical, Organizational and Leadership Questions of the Hungarian History of Higher Officer Training], *Akadémiai Közlemények* 178 (1991): 20; Dr. Antal Oroszi, *A Zrínyi Miklós Katonai Akadémia történetének összefoglalása (1947–1996). A magyar katonai vezető- és tisztképzés története (Tanulmánygyűjtemény)* [Summary of the History of the Miklós Zrínyi Military Academy, 1947–1996. The History of the Hungarian Military Leadership and Officer Training (Collection of Studies)] (Budapest, 1996), p. 253.

11. HM HL HAKAD 01723/H. A.–1953 (Nov. 25).

12. *Ibid.*
13. HM HL HAKAD 02996/H. A.–1952 (Dec. 3).
14. HM HL HAKAD 01723/H. A.–1953 (Nov. 25).
15. HM HL HAKAD 02996/H. A.–1952 (Dec. 3).
16. Bérczes rose to be major general and commander of the Tata armored division.
17. HM HL HAKAD 05/H. A.–1953 (Jan. 6).
18. HM HL HAKAD 03388/H. A.–1951 (Dec. 5).
19. HM HL HAKAD 03120/H. A.–1952 (Dec. 10).
20. HM HL HAKAD 05/H. A.–1953 (Jan. 6).
21. HM HL HAKAD 03149/H. A.–1952 (Dec. 14) and 03261/H. A.–1952 (Dec. 24).
22. HM HL HAKAD 01723/H. A.–1953 (Nov. 25).
23. HM HL HAKAD 03326/H. A.–1952 (Dec. 16).
24. HM HL HAKAD 03208/H. A.–1952 (Dec. 18).
25. HM HL HAKAD 03311/H. A.–1952 (Dec. 27).
26. HM HL HAKAD 03334/H. A.–1952 (Dec. 30).
27. HM HL HAKAD 073/H. A.–1953 (Jan. 20).
28. HM HL HAKAD 083/H. A.–1953 (Jan. 15).
29. HM HL HAKAD 0155/H. A.–1953 (Feb. 5).
30. HM HL HAKAD 0299/H. A.–1953 (March 5).
31. HM HL HAKAD 0511/H. A.–1953 (April 14).
32. HM HL HAKAD 0807/H. A.–1953 (May 23).
33. See Note 27.
34. HM HL HAKAD 0872/H. A.–1953 (June 25).
35. *Ibid.*
36. HM HL HAKAD 001043/H. A.–1953 (Aug. 5).
37. The Central Committee of the Hungarian Workers' Party, meeting on June 27–28, passed a resolution recalling General Mihály Farkas, minister of defense. Parliament, also on the Party's proposal, elected Lieutenant General István Bata as minister of defense on July 4. So this report was dealt with mainly by Major General Béla Székely, appointed as chief of General Staff on August 10, 1953.
38. HM HL HAKAD 01088/H. A.–1953 (Aug. 22).
39. HM HL HAKAD 01120/H. A.–1953 (Sept. 1).
40. HM HL HAKAD 01122/H. A.–1953 (Sept. 1).
41. HM HL HAKAD 01158/H. A.–1953 (Sept. 8).
42. See Note 35.
43. HM HL HAKAD 001204/H. A.–1953 (Sept. 25).

44. Only first and second years were held that academic year; the full strength was reached with the addition of the third year in the following academic year.
45. HM HL HAKAD 001400/H. A.–1953 (Oct. 17).
46. HM HL HAKAD 001558/H. A.–1953 (Nov. 4).
47. HM HL MN 1953/T. 652/2. cs.–01682/H. A. Be.–1953.
48. For the most important of these, see Notes 23, 27, 32 and 35.
49. See the relevant sections of Notes 23, 27 and 32.
50. HM HL HAKAD 01723/ H. A.–1953 (Nov. 25).
51. To the 383 Hungarian People's Army students mentioned earlier there must be added 40 ÁVH students, among whom, it seems, there was also some wastage.

Chapter VIII
The Final Year and a Half of the National Defense Academy
(November 1, 1953–March 14, 1955)

The bare year and a half indicated in the title was a highly ambivalent period in foreign and in domestic policy. The steady normalization in 1953–1954 of Yugoslavia's relations with the Soviet Union and the Socialist countries had a positive effect on the international climate. So did the Berlin talks on January 25–February 18, 1954, between the American, British, French and Soviet foreign ministers, the defeat of the French at Dien-bien-phu on May 7, 1954 (by creating conditions in which the war in Indochina could be ended), the Zhou Enlai–Nehru declaration in New Delhi of the five principles of peaceful coexistence on June 28, 1954, the Indochina agreement signed in Geneva on July 21 that year, and the January 25, 1955, order of the Presidium of the Supreme Soviet ending the state of war between the Soviet Union and Germany.

But the Eastern bloc saw threats, for instance, in the US presidential approval on August 24, 1954, of the Communist Control Act, the foundation of the Southeast Asia Treaty Organization (SEATO) on September 8, and the admission of West Germany into NATO at the nine-power conference in London on September 28–October 3. The Soviet response to the last event was the Moscow "Conference of European Socialist Countries on European Peace and Security" on November 29–December 2. The October 12 confirmation of cooperation between the Chinese People's Republic and the Soviet Union was followed on December 2 by a security agreement between the United States and Chiang Kai-shek's Taiwan. In November 1954, armed struggle against France broke out in Algeria. On December 30, the French National Assembly voted through the Paris agreements that enabled West

German rearmament and admission to NATO. In February 1955, the Baghdad Pact brought into being a treaty organization for the Middle East, known as CENTO.

There was similar variety in domestic politics and events in the People's Army. The standing of military higher education was enhanced by a government order (HT VIII. 8) granting the National Defense Academy (and the Stalin Academy) the status of a college,[1] for which the general educational requirement for entrance was *temporarily* to be a specialist school-leaving certificate. But this positive development was offset by the uncertainty of an officer's career in terms of livelihood: the strength of the Hungarian army was reduced on November 15, 1953, to 151,000, and 2,471 officers were made redundant.[2]

Up to Stalin's death on March 5, 1953, the Soviets had never officially stated that the Eastern bloc would not attack Yugoslavia. After Stalin's death, the processes of reconciliation with Tito and reduction in the bloc's armed forces could begin. This lasted up to 1956, with annual reductions in the armed forces being accompanied by radical cuts in the officer corps as well. Those who remained had to change posts almost every year, and every officer felt that his situation was insecure, and so the People's Army in 1956 was led by disillusioned officers. At the same time, there began a phase of qualitative development in the army, as necessitated by the appearance of the atom bomb and other weapons of mass destruction. The degree of mechanization and instrumentation in the Hungarian People's Army was enhanced and the armored forces were increased.

In the spring of 1954, the Central Committee of the Hungarian Communist Party expressed concern about the loosening of discipline among army officers, and, seizing upon possible contributing factors, urged that their housing problems be solved and that they be relocated less frequently. At the Third Congress of the Party on May 24–30, Mátyás Rákosi heaped praise on the People's Army, but widespread uncertainty among soldiers resulted from the Congress decision to bring Mihály Farkas back into the Politburo and the Central Committee Secretariat—the soldiers had not been given real reasons for his earlier dismissal as minister of defense in any case. Another development running against the tide of reductions—although serving as a basis for higher quality in the officer corps—was the fact that the Ferenc Rákóczi II Military High School passed to a new organization in July 1954 and had its student roll increased from 200 to 600.[3]

A government resolution of September 10, 1954, stated that those demobilized by the arms reductions should be assisted to integrate into society, and called on ministries to give them preference when recruiting staff. They were entitled to one month's pay and clothing allowance. It was decided at a meeting of higher commanders of the Hungarian People's Army on November 1 to introduce organizational, operational and combat principles for the use of nuclear weapons and to develop infantry units into motorized infantry. New Combat Regulations that took the use of nuclear weaponry into account came into force on December 1.

The Military Council of the Ministry of Defense established at a meeting on December 11, 1954, that the Hungarian People's Army had 4 major generals, 50 colonels, 154 lieutenant colonels and 225 majors serving who had also been members of the Royal Hungarian Defense Force. One of these held the position of corps commander, 6 that of regimental commander, and 24 that of battalion commander, while 3 others held staff positions in a corps and 6 in a division. As for the social situation of officers, 6,500 were found to have no apartment, partly because 5,600 officers had been posted to different positions in 1953–1954.[4]

The Military Council concurrently sent the Politburo a report on the cadre situation in 1954. The officer corps' social composition had improved, with 52.34 percent of worker origin and 26.02 percent from the peasantry; 65 percent were Party members and

4.78 percent had served in the Horthy period; 85 percent had completed eight grades of elementary education and only 15 percent had graduated from high school. It was reported that 40 percent of the medical officers were requesting their dismissal; this was mainly for financial reasons, but they were also dissatisfied at not being able to obtain specialist medical training.[5]And the report did not cover the data of December 31, by which time the strength of the Hungarian People's Army had declined by 151,265 and 2,843 officers had been dismissed over a period of a year.[6]

The main foreign and domestic policy relations that were reflected and the state of the Hungarian People's Army were the basic determinants of the National Defense Academy's activity in 1953–1955. In the last third of November 1954, the commander of the Academy, Colonel Mihály Gyulai, had to submit the reports of the faculties on the start of the academic year.[7]

It emerges from the All-branch Faculty report that A (first-year), B (second-year), and T (extension training) groups consisting of four classes of 72, two classes of 32 and two classes of 33 students—a total roll of 137, including the 20-man A–11 supplies class—had begun their studies. In their previous education, 29 percent had completed five or six years of grade school, 35 percent seven or eight years, and 29 percent middle school; just 7 percent had graduated from high school. In social composition, 48 percent came from a worker background, 44 percent from the peasantry, 3 percent from the intelligentsia, and 5 percent from other employed strata. The class grade averages were in the 3.1–3.9 range and the average for the faculty was 3.5. As in previous years, it was found that first-year students had first to be taught how to study.[8]

The Engineering and Signals Service Faculty report[9] spoke of one engineers' and one signals troops' class each in the first and second years, of 17 and 15, and 17 and 13 men respectively, a total of 62 students, all of whom had completed the all-branch officers' school. In civilian education, 23 percent had completed six years of elementary school, 49 percent eight years of elementary school or four years of civil school, and 14 percent from five to seven years of high school or two years of commercial school, while 14 percent had graduated from high school. The class grade averages ranged from 3.69 to 3.86, with a course average of 3.75. Performance was weaker in general tactics, arithmetic, Russian language and social studies. Relations between the faculty commander and the specialist departments were exemplary.

The commander of the Academy, in an order dated November 27, 1953,[10] made an assessment of the 1952–1953 academic year, and took steps to implement the minister of defense's order for the 1953–1954 academic year. He considered that the National Defense Academy had basically performed its tasks in the 1952–1953 year; the study results had improved over the previous year as well.

An important change at the National Defense Academy came on December 1, 1953, when the minister of defense appointed Major General Endre Matékovits as commander of the Academy.[11]

On January 2, 1954, the commander submitted a report on changing the program for the Correspondence Course to the chief of General Staff, to ensure that students would hear the main lectures on general and branch combat, politics and military history, and be able to consult lecturers on these.[12] At about this time the Topography/Military Geography Department underwent a qualitative change in the syllabus and lectures on military geography. Great care was taken to build up the practical side of topography training; 70 percent of the work was to be done in the field.[13]

The Academy submitted a proposal to the minister of defense on August 24, 1954,[14] that the name of the National Defense Academy (*Honvéd Akadémia*) be changed from the beginning of the 1954–1955 academic year to the Miklós Zrínyi Military Academy (*Zrínyi Miklós Katonai Akadémia*). According to one paragraph in the thorough and scholarly argument for this, "It will be 390 years on November 18 this year since Miklós Zrínyi—under circumstances still not fully clarified to this day—died a martyr's death for his unflinching principles, by the weapon of a hired assassin of the Vienna court. At a well-prepared ceremony on this day, the Comrade Minister of Defense could inform the complement of the Academy of the fact of the name change, which will undoubtedly remain forever in the memories of the officers serving here and spur them to produce still better work." The essence of the whole proposal was the idea that "Miklós Zrínyi, through his excellent qualities, became a great politician, an unswerving warrior for the idea of an independent Hungary, the greatest poet of his age, a many-sided military scholar… [and] a military commander reliant on the people and therefore successful. In view of all this, his figure is eminently suitable for placing before the officers of the Academy as a paragon… Zrínyi, an example not only of active patriotism but of the inseparable unity of theory and practice, will teach [a lesson to] our officers in how to defend this country of ours, which is building socialism."

On October 5 the commander of the Academy submitted a proposal for issuing an order on the 1954–1955 training year for the Correspondence Faculty, requesting that the chief of General Staff standardize the number of study days for the correspondence students: every Monday for the Budapest students, every Wednesday for the Transdanubians, and every Friday for those from the Danube–Tisza and Trans-Tisza regions. It would only be possible to alter all this with the consent of the commander of the Correspondence Faculty. There would be six days' study leave due before each half-yearly conference. Correspondence students would be relieved of high school and political studies. He recommended that the chief of General Staff's order declare an inescapable obligation to attend the conferences and field exercises. Despite expectations, the Correspondence Faculty still did not operate smoothly. Major General Matékovits mentioned this in another memo to the chief of General Staff,[15] reporting that 12 students had failed to sit the end-of-year examinations for the first year.

The commander sent a report on the situation at the National Defense Academy between April 1 and September 30. The Academy Command felt that there had been a slow but steady improvement in the disciplinary situation and morale, but they had still not reached the desired standard. During the period examined, 140 students (42 percent of the roll) had received reprimands, ranging from rebukes, through 20 days' detention in the garrison jail, to demotion. The Court of Honor had expelled three persons. On the other hand, 172 students (50.6 percent) had received commendations. Of the staff and teaching staff, 38 (15 percent) had received altogether 38 reprimands, ranging from rebukes to 20 days in the garrison jail, while 137 (57 percent) had received a total of 179 commendations. Half the reprimands in each section of the complement had been for lateness, but there were plenty of problems with immoral life and immoderate drinking as well. According to the Academy Command, the disciplinary situation had not improved to the desired and expected standard during the academic year.

The study situation appears in the following examination results:

Year/Course	Study Average	Excellent	Good	Satisfactory	Not Satisfactory
First Year	3.94	18	101	35	25
Second Year	3.93	15	80	33	4
Extension Course	4.13	8	35	8	-
Whole Academy	*4.00*	*41*	*216*	*76*	*29*

The first-year and second-year Academy Courses showed fairly similar dispersions of performance, except in the "Not Satisfactory" column. The Extension Course did better than it had in the previous year in every respect. The class averages ranged between 3.81 and 4.06 in the first year and 3.81 and 4.06 in the second. The commander of the Academy noted a significant qualitative change in the teaching staff, in knowledge and in methodology, especially in the Artillery, Signals Troops, Engineering and Armored Faculties. He regretted to conclude that General Combat had still not developed into the Academy's foremost faculty, mainly, in his view, because the otherwise talented, promising young faculty commander lacked group or other practical experience, being simply a man of theory. He found the Anti-aircraft and Air Force Faculties weakest: "The heads of these faculties do not reach the Academy standard; their teaching staffs are weaker compared with the other faculties."

The report includes for the first time in the Academy's history a thorough, almost sociological analysis of the teaching staff. This reveals that in their previous education, 6 had obtained a Soviet academy diploma, 12 had completed a Soviet course, 27 had obtained a two-year National Defense Academy diploma, 1 had completed the Academy's one-year Higher Commander Course, 53 had completed the Academy's Regimental Commander Course, 20 had passed out from the Ludovika Military Academy, and 41 had acquired an education that did not include a relevant qualification. Thus 18.3 percent had a diploma, 47.7 percent had completed a course, 11.2 percent had completely military higher education in the Royal Hungarian Defense Force, and 22.8 percent had no specialist training. Only the first of those categories could be considered as adequate for providing up-to-date Academy instruction. Nor was the picture of civilian educational attainment more reassuring: 19 (10.5 percent) with a higher education, 58 (32.3 percent) with a high school certificate, 78 (43.3 percent) who had completed a four-year middle school course, and 5 (2.7 percent) with less education than that. Finally, there was trouble on the level of field or practical experience: 54 (30 percent) had none, 8 (4.4 percent) had 6 months, 47 (26.2 percent) had a year, 46 (25.5 percent) had 2–5 years, 11 (6.2 percent) had 5–10 years, and only 2 (1.1 percent) had over 20 years' practical experience. This was somewhat offset by the fact that the General Staff had provided 90 persons with a total of 2,057 days of supervisory review and army, corps and division staff exercises during the academic year. This meant that 50 percent of the teachers had spent an average of 23 days with the troops on such occasions.

Major General Matékovits reported in detail on the preparations and tasks for the 1954–1955 academic year. The chief of General Staff had apparently endorsed instruction plans based on new regulations (except for the Engineering Faculty and Correspondence Faculty) laying down the following thematic proportions: social science studies 16 percent, operational and tactical training and specialist training 60 percent, and general studies 24 percent.

October that year was declared a methodology month, to be spent partly on central and partly on faculty methodological activities. The methodological preparations were centered on study of the new regulations. As part of the preparations for the year, first-year engineers, signals troops and anti-aircraft students began on September 20, 1954, to study mathematics, physics, Hungarian language and topography. The Military History Faculty prepared notes that allowed the teaching of military history to be uniform. There also appeared a textbook for the new subject of military geography. The commander of the Academy proposed to his superiors that the training period in the Engineering Faculty be increased from three years to four. To increase the proportion of graduate teachers, it was planned that 10 or 15 officers should join the first-year Correspondence Course and a smaller number the three-year Academy course. It was intended that the vast majority of the teaching staff should have an Academy qualification by 1960–1961.

In the 1954–1955 academic year, there were 160 students in each year of the three-year Academy course, 40 on the Extension Course, and 52 and 58 students on the two years of the Correspondence Course, making a total of 630 students, to which were added 15 students sent by the secret police, the State Security Authority (ÁVH). This roll called for 35 classes, discounting those of the Correspondence Faculty. There also began a special (Party functionary) course of 40–45, and a 6-man special group under the command of Major General Szabó, within the framework of the Correspondence Faculty. This meant that discounting the two special groups, the student roll had increased by 57.5 percent since the previous year. Meanwhile, the number of teachers had risen from 183 to 222 (by 17.1 percent) and the administrative staff had been reduced from 614 to 545 (by 11.1 percent). The number of students per ancillary staff member had fallen from 1.99 to 1.2 and per teacher from 2.1 to 2.8.

The commander's orders brought some tangible results. For example, officers and young engineers who had completed a two- or three-year Academy course were recruited for the Signals Troops Faculty,[16] which steadily remedied the weakness of having some teachers with a very low level of operational and tactical knowledge. For the Arms, Armored and Vehicle Mechanics Faculty, it brought a marked advance when the teaching of armored and vehicle mechanics was extended in the 1954–1955 training year to all branch and specialist officers—that is, when it became general in the Academy.[17] But there seems to have been a lack of foresight in the fact that first-year classes that were started for navigating officers and for staff officers in 1953–1954 were then merged in 1954–1955.[18]

Colonel General István Bata, the minister of defense, announced the following in his order of March 14, 1955:[19] "On my proposal, the Council of Ministers of the Hungarian People's Republic, with effect from March 15, 1955, has granted the National Defense Academy the honorable name of MIKLÓS

ZRÍNYI MILITARY ACADEMY." This ended a very important phase in Hungarian higher military education after World War II. After the brief operation of the National Defense War Academy and its *de facto* discontinuation, higher Academy training for officers was reborn in this institution over the 54 often extremely difficult months (exactly four and a half years to the day) between September 14, 1950, and March 14, 1955.

The National Defense Academy was the Hungarian People's Army's cadre-training institution, but it has been seen that in its commanders, Major General Béla Király, Colonel Mihály Gyulai and Major General Endre Matékovits, as well as teachers such as Colonel Imre Lóránt, Lieutenant Colonels Sándor Zsilinszky and Ákos Vizi-Makkay, Colonels Imre Homér, Pál Naményi and Géza Buzsáki, and many others, it had, even in the autumn of 1954, teachers who had received their officer training, knowledge, expertise and devotion to the cause in the Royal Hungarian Defense Force, and provided the cause of modernization with continuity between the pre-war War Academy and the National Defense Academy.

New leaders possessing quite another ideological basis, such as Colonel Lajos Tóth and several other staff and teaching officers at the Academy, worked to some extent in conjunction with the officers from the "old army" to continue their activity and turn out officers trained to as high a level as possible, to defend the home country, without any qualifying adjectives. Earlier accounts and the statistics on previous pages show what heroic work had to be done to overcome the problems of general education, professional skills, military experience, and so on, and steadily perfect the structure, methodology and overall preparedness of the officers graduating from the institution. Throughout the review of the National Defense Academy's activity over less than five years, this constant effort by the Academy Command to aim at a truly modern Academy training regime, and to achieve this where circumstances allowed, was apparent.

The Composition of the Military Men

Ser. no.	Name		General and officer	Sub-officer		Hon-véd	Student at military school	Total no. of soldiers
				Re-enlisted	Regular			
			According to party affiliation					
1.	Party member		686	19	5	3	-	713
2.	Candidate for membership		14	2	1	1	-	18
3.	Member of DISZ*		8	17	7	34	-	66
4.	Outside party or DISZ organization		57	18	5	58	-	138
5.	Total:		765	56	18	96	-	935
			According to social extraction					
6.	Worker		278	23	6	24	-	331
7.	Peasant		244	24	12	60	-	340
8.	Intellectual		26	5	-	1	-	32
9.	Employee		125	3	-	2	-	130
10.	Artisan, retailer		67	1	-	9	-	77
11.	Other		25	-	-	-	-	25
12.	Total:		765	56	18	96	-	935
			According to profession					
13.	Worker		357	34	13	42	-	331
14.	Peasant		165	11	4	48	-	34
15.	Intellectual		9	-	-	-	-	32
16.	Employee		125	8	1	5	-	130
17.	Student		104	3	-	1	-	77
18.	Other		5	-	-	-	-	25
19.	Total:		765	56	18	96	-	935
			According to educational level					
20.	Without education	Illiterate	-	-	-	2	-	2
21.		Read and write	-	-	-	-	-	-
22.	1–4 elementary school		2	3	1	27	-	33
23.	5–8 elementary, 1–4 secondary school		429	43	15	65	-	552
24.	5–6 secondary school		99	1	1	-	-	100
25.	7–8 secondary school		67	3	-	-	-	71
26.	Maturity		118	6	1	2	-	127
27.	College		37	-	-	-	-	37
28.	University		13	-	-	-	-	13
29.	Total:		765	56	18	96	-	935

* DISZ is Democratic Youth Alliance

The Composition of the Military Men (cont.)

Ser. no.	Name		General and officer	Sub-officer		*Hon-véd*	Student at military school	Total no. of soldiers
				Re-enlisted	Regular			
	According to military education							
30.	Academy, officers' school		457	-	-	-	-	457
31.	Course		289	31	9	-	-	329
32.	Trained by practice		1	25	9	42	-	77
33.	Untrained		18	-	-	54	-	72
34.		Total	765	56	18	96	-	935
	According to fitness for service							
35.	Fit for armed service		753	53	18	94	-	918
36.	Fit for special qualifi-cation	With physical restrictions	12	1	-	-	-	13
37.		Without physical restrictions	-	2	-	2	-	4
38.		Total:	765	56	18	96	-	935
	According to marital status							
39.	Married		573	38	5	14	-	630
40.	Divorced (widow/er)		10	1	1	-	-	12
41.	Single		182	17	12	82	-	293
42.		Total:	765	56	18	96	-	935
	According to year of birth							
43.	1937 and later		-	-	-	-	-	-
44.	1936		-	-	-	-	-	-
45.	1935		-	-	-	-	-	-
46.	1934		1	1	-	49	-	51
47.	1933		9	3	11	33	-	56
48.	1932		31	4	6	10	-	51
49.	1931		84	10	1	4	-	99
50.	1930		112	10	-	-	-	122
51.	1929-1925		325	15	-	-	-	340
52.	1924-1920		109	2	-	-	-	111
53.	1919-1915		45	4	-	-	-	49
54.	1914-1910		33	6	-	-	-	39
55.	1909-1905		10	1	-	-	-	11
56.	1904-1900		2	-	-	-	-	2
57.	1899-1895		3	-	-	-	-	3
58.	1894 and before		1	-	-	-	-	1
59		Total:	765	56	18	96	-	935

The Composition of the Military Men (cont.)

Ser. no.	Name		General and officer	Sub-officer		Hon-véd	Student at military school	Total number of soldiers
				Re-enlisted	Regular			
According to years of military service								
60.	1 year			-	-	54	-	54
61.	2 years		2	-	-	38	-	40
62.	3 years		3	2	12	4	-	21
63.	4 years		50	7	6	-	-	63
64.	5 years		87	21	-	-	-	108
65.	6–10 years		519	15	-	-	-	534
66.	11–15 years		52	2	-	-	-	54
67.	16–20 years		22	6	-	-	-	28
68.	21–25 years		15	1	-	-	-	16
69.	26–30 years		8	2	-	-	-	10
70.	31–40 years		5	-	-	-	-	5
71.	Over 40 years		2	-	-	-	-	2
72.	Total:		765	56	18	96	-	935
According to knowledge of a language								
73.	Slav	Russian	50	-	-	-	-	50
74.		Other Slav	24	1	-	-	-	25
75.	German		37	-	-	-	-	37
76.	English		2	-	-	-	-	2
77.	French		6	-	-	-	-	6
78.	Italian		3	-	-	-	-	3
79.	Other			-	-	-	-	-
80.	Without knowledge of a foreign language		663	55	18	96	-	832
81.	Total:		785	56	18	96	-	955

The Composition of the Military Men (cont.)

Ser. no.	Name	Staff no.
	According to rank	
82.	Colonel-general	-
83.	Lieutenant general	-
84.	Major general	1
85.	Colonel	12
86.	Lieutenant-colonel	46
87.	Major	70
88.	Captain	134
89.	First lieutenant	299
90.	Lieutenant	202
91.	Sub-lieutenant	1
92.	**Total of generals and officers**	**765**
93.	Company sergeant-major	7
94.	Sergeang first class	28
95.	Sergeant	15
96.	Lance sergeant	4
97.	Corporal	2
98.	**Total of re-enlisted sub-officers**	**56**
99.	Sergeant first class	-
100.	Sergeant	2
101.	Lance sergeant	8
102.	Corporal	8
103.	**Total of regular sub-officers**	**18**
104.	Sergeant first class	-
105.	Sergeant	-
106.	Lance sergeant	-
107.	Corporal	-
108.	Private soldier	-
109.	**Total of students**	**-**
110.	Private first class	6
111.	Honvéd	90
112.	**Total of honvéds**	**96**

According to the acceptance of the re-enlisted years							
The re-enlisted sub-officer			The actual year of the accepted time				Total
			1.	2.	3.	4.	
113.	Accepted years	2	2	7	-	-	9
114.		3	-	-	-	-	-
115.		4	16	8	6	17	47
116.	Total		18	15	6	17	56

The Composition of the Military Men (cont.)

Ser. no.	Name	General and officer	Re-enlisted sub-officer	Regular sub-officer		Honvéd		
				2.	3.	1.	2.	3.
				years of serving time				
According to more important professions, qualifications								
117.	Coal-miner	13	-	-	-	-	-	-
118.	Other miner	4	-	-	-	-	-	-
119.	Iron turner	18	3	-	1	-	1	-
120.	Motor mechanic	6	-	-	-	1	-	-
121.	Locksmith	48	5	2	1	1	4	3
122.	Mechanican	31	-	-	-	-	-	-
123.	Electrician	7	2	-	-	-	-	-
124.	Bricklayer	8	1	1	-	4	1	-
125.	Joiner, carpenter	15	-	-	-	-	-	-
126.	Saddler	-	-	-	-	-	-	-
127.	Vulcanizer	-	-	-	-	-	-	-
128.	Chemist	-	1	-	-	-	-	-
129.	Tailor, furrier	19	2	1	-	1	1	-
130.	Shoemaker	11	-	-	-	-	-	-
131.	Baker, confectioner	4	-	-	-	-	-	-
132.	Cook, butcher	3	-	-	-	-	-	-
133.	Driver	2	3	1	-	-	-	-
134.	Tractor-driver	2	-	-	4	6	1	-
135.	Photogapher, optician	2	-	-	-	-	-	-
136.	Agricultural laborer	158	17	3	-	28	17	-
137.	Physician	2	-	-	-	-	-	-
138.	Engineer	-	-	-	-	-	-	-
139.	Pharmacist	-	-	-	-	-	-	-
140.	Teacher	6	-	-	-	-	-	-
141.	Student	106	2	-	-	-	-	-
142.	Technician	4	1	-	-	-	1	-
143.	Draughtsman	-	-	-	-	-	-	-
144.	Telegraphist	-	-	-	-	-	-	-
145.	Engine-driver, fireman	-	-	-	-	-	-	-
146.	Sailor	-	-	-	-	-	-	-
147.	Statistician	-	-	-	-	-	-	-
148.	Nurse	-	-	-	-	-	-	-
149.	Musician	1	-	-	-	-	-	-
150.	Other	304	19	4	-	13	12	1
151.	Total:	765	56	12	6	54	38	4

Reports
I report that the surplus appearing at the distribution according to the knowledge of a language is caused by the fact that there are several officers at the Military Academy who can speak more than one languages, so I noted each of them at all the proper languages.

The last set of statistics showed a marked advance in the educational attainment, military preparedness and linguistic knowledge of the officers. Also encouraging was Major General Matékovits's proposal to educate the teaching staff "at a rapid rate." All in all, in the author's view, the huge effort by commanders, staff officers and teachers of the National Defense Academy had created a modern institute of higher military education and laid the future basis for the Miklós Zrínyi Military Academy.

Notes

1. *A magyar állam szervei 1950–1970. Központi szervek* [Bodies of the Hungarian State, 1950–1970. Central Bodies] (Budapest, 1993), p. 372.

2. László Csendes and Tibor Gellért, *Kronológia a Honvédség történetéből 1945–1990* [Chronology of the History of the Honvédség, 1945–1990] (Budapest, 1993), p. 136.

3. *Ibid.*, p. 139.

4. László Csendes and Tibor Gellért, *Háborútól a forradalomig 1945–1955. Adalékok a magyar hadsereg történetéből* [From War to Revolution, 1945–1955. Contributions to Hungarian Army History] (Budapest, 1994), p. 78.

5. *Ibid.*, p. 82.

6. Csendes and Gellért, *Kronológia a Honvédség történetéből 1945–1990*, p. 143.

7. Honvédelmi Minisztérium, Hadtörténelmi Levéltár, Honvéd Akadémia [Ministry of Defense, Military History Archive, National Defense Academy] (hereafter cited as HM HL HAKAD) 01727/H. A.–1953 (Nov. 26).

8. HM HL HAKAD 01731/H. A.–1953 (Nov. 22).

9. *Ibid.*, 01734/H. A.–1953 (Nov. 22).

10. *Ibid.*, 001944/H. A.–1953 (Nov. 4).

11. See the Biographies of Key Personalities.

12. HM HL HAKAD 033/H. A.–1954 (Jan. 2).

13. Lieutenant General Imre Gábor, "Tereptan—katonaföldrajzi tanszék" [The Topography/Military Geography Department], *Akadémiai Közlemények. Az akadémia szerveinek története III/B kötet. A Szárazföldi ágazat tanszékeinek története* 120 (1985) 3: 192.

14. HM HL HAKAD 01982/H. A.–1954 (Aug. 25).

15. *Ibid.* 02416/H. A.–1954 (Oct. 20).

16. Colonel Konstantin Déry and Lieutenant Colonel János Estók, "Híradó tanszék" [The Signals Department], *Akadémiai Közlemények* 120 (1985) 3: 9.

17. Dr. Lieutenant Colonel István Rajkó, "Fegyverzeti-, páncélos- és gépjárműtechnikai tanszék" [The Arms, Armor and Motorization Department], *Akadémiai Közlemények* 120 (1985) 3: 98.

18. Dr. Lieutenant Colonel Jenő Bartos, Lieutenant Colonel Lajos Kovács (ret.) and Lieutenant Colonel József Hajdó (ret.): "Repülő hadműveleti–harcászati tanszék" [Air Force Operational Tactical Faculty], *Akadémiai Közlemények. Az akadémia szerveinek története IV kötet. A légvédelmi és repülő ágazat tanszékeinek története* 120 (1986) 4: 22, 25.

19. According to the research of Colonel Dr. Antal Oroszi (ret.).

Chapter IX

Higher-Level Training for Political Officers in Hungary
(September 1, 1948–December 16, 1956)

The Hungarian Communist Party after World War II worked steadily to gain ground in the new democratic army. Among the first of a succession of measures to that end came in March 1946, when the National Defense Education Officers' Institution was formed out of reliable cadres selected from among the members of all democratic parties. To provide them with as thorough a training as possible, an Education Officers' Course was started in October that year, teaching political subjects that were intended to assist them in their work.

Understanding that period of Hungarian history calls for clear insight into the domestic developments and the essence of Hungary's international relations. For both were controlled by the Soviet Union. The absolute power over Hungary held since the liberation by the Allied Control Commission ended on September 15, 1947, when the Treaty of Paris came into force. By that time, the Hungarian Communist Party, encouraged by the Soviet-dominated Allied Control Commission, was destroying the infant democratic political and social system, using a series of "salami tactics," and establishing a Soviet-style one-party state. On July 1, 1948, the Social Democratic Party, the long-established political wing of the labor movement, was merged with the Communists into the Hungarian Workers' Party (MDP)—a Bolshevik Party in all but name. The subsequent dictatorship by Mátyás Rákosi and a tiny clique imposed classic Stalinist terror and total control over every aspect of society and the economy. The death of Stalin and subsequent changes in Moscow allowed Imre Nagy's first government of 1953–1955 to try to reform the system through a New Course of "communism with a human face," but the Bolshevik wing of the

Party struck back in 1955, only to be overtaken by a wave of reform sentiment and grassroots discontent that culminated in the Hungarian Revolution of 1956 under Imre Nagy's leadership. But peaceful efforts at reform were met by Soviet-led Bolshevik force, and the Hungarian bid for freedom and independence lasted only a few days. Then came a massive exodus of refugees and fierce reprisals by the Soviet-sponsored regime of János Kádár. This succession of events provided the framework against which political officers were being trained for the army in that period.

1948 was also a turning point in foreign relations, for Tito's Yugoslavia, hitherto a trusted ally of the Soviet Union, suddenly fell foul of Stalin. Long a victim of megalomania, Stalin would brook no independent action by Communist leaders in satellite countries, but Tito, whose war record gave him greater domestic legitimacy than the others, wished to take communism in Yugoslavia down a different road from the Soviet Union's. On June 27, 1948, Stalin ordered the Yugoslav Communist Party's expulsion from COMINFORM, the Soviet-run Communist Party umbrella organization. Stalin was considering all methods of removing Tito, including war, for which rapid preparations were made. This also meant that satellites such as Romania, Bulgaria and Hungary had to build up larger military forces than the post-war peace treaties allowed them. Theirs became war economies, at the expense of living standards, for example. By 1950, they were ready. The United States and the West up to then had been seen as "paper tigers," but this assumption was disproved by events in the Korean War. Had it not been, the next victim would have been Yugoslavia. Not that this could have been openly divulged while Stalin was alive. He died, however, on March 5, 1953, and although the propaganda war continued for a time, efforts at reconciliation began and annual reductions in Soviet-bloc arms levels became possible. These cutbacks caused fears for their livelihood to spread among officers of the Hungarian People's Army.

Those are the events against which the political officer corps—an organization of exclusively Communist persons, replacing the multi-party system of education officers—needs to be studied. The provision of such officers throughout the armed forces called for an immense investment of money and manpower. The political officer of each unit counted as its joint commander from February 19, 1949, until January 20, 1953, when they were redefined as the commander's deputy. In addition to this change, in the authority of the political officers, their selection and training showed three characteristics:

- Social origin was a more decisive selection factor than knowledge or talent.

- Their overwhelming majority had extremely low levels of general education. Even in the autumn of 1955, only half the students in the highest echelon of political officers' training had a certificate of secondary education (*érettségi*).

- Loyalty to Stalin was paramount. Political officer candidates had to be servile.

On September 9, 1948, Mihály Farkas, hitherto the deputy general secretary of the Hungarian Workers' Party, became minister of defense. His "Pilis II" chart of organization on October 1 established the Petőfi National Defense Academy for Education Officers, which had started its first five-month course on September 1. Its task was to provide higher-level training for political officers. Soviet experience and advice suggested that a decisive role during the interim period of building up a socialist-style army had to go to political officers, working alongside commanders. Such political officers had decisive sway over political education.

On November 29, 1948, the National Conference of Education Officers convened to discuss the underlying principles of political education, summed up in the following slogan: "You're responsible for the whole unit."[1] The new minister of defense announced that a system of political officers would be introduced. As a first step, an order of December 3, 1948, on reorganization of the leading bodies of the Ministry of Defense established were the National Defense General Staff, the Personnel Section, and the Branch Inspectorates (six months later renamed Commands) a Political Section (six months later a Political Chief Section) headed by Lieutenant General Sándor Nógrádi. This turned the system of command over the National Defense Force into a replica of the Soviet system.

Just nine days later, on December 12, the Petőfi National Defense Academy for Education Officers opened[2] at Vöröshad-sereg útja 21–23 in Budapest's Second District (the former Bolyai Engineering Academy of the Royal Hungarian Defense Force), under the command of Colonel Dr. Dénes Felkai.[3] The ceremony was attended by Mátyás Rákosi, the general secretary of the Communist Party, several other Party, state and military leaders, and representatives of the Soviet advisory group. At the ceremony, representatives of the Red Csepel industrial complex handed over to the Academy its unit banner (with the legend "With the people through fire and water") and a silver cornet.[4] Of the first-year students, 60 percent came from a worker background and 38 percent from the peasantry.[5] They were sworn in as political officers on February 11, 1949.[6]

The minister of defense, in an order dated February 15, 1949, changed the name of the institution from the Petőfi National Defense Academy for Education Officers to simply the Petőfi National Defense Academy.[7] The November announcement was followed by a resolution of the Communist Party Central Committee and a Ministry of Defense measure on February 18, 1949, introducing the system of political officers on March 15, 1949, based on the "Klapka" chart of organization. It meant that the National Defense Institute of Education Officers had completed its work, or continued it under the aegis of the National Defense Institute of Political Officers.[8] According to the Ministry of

Defense measure, the task of the political officers was "to conduct militant Marxist-Leninist political education that promotes to the full the introduction of modern training and readiness for battle, and a uniform spirit, discipline and order, in other words strike capability, in the National Defense Force."[9] Political officers accordingly received extremely wide powers as co-commanders. Orders were valid only if signed by both the commander and the political officer. If their opinions differed, the procedure was for them to consult both a higher officer and a political officer, but if that was not possible the political officer had the right and obligation to issue the command under his own signature, which the subordinates were obliged to obey, although the political officer had to take responsibility for it. In line with the political officer's political function, he might be consulted without going through official channels. But the ministerial order emphasized the point that the commanders and political officers had to work for comradely cooperation, in the interests of combat preparedness and the moral-cum-political education of the troops.

On March 1, 1949, Lieutenant General Sándor Nógrádi, deputy minister of defense, was appointed to head the Political Chief Section that ran the political work of the National Defense Force. Five days later, a further 38 political officers were sworn in at the Petőfi National Defense Academy, based on Party committee recommendations. They and their comrades-in-arms had to wear, from April 26 onwards, a distinguishing mark on the sleeves of their tunics and greatcoats: a five-pointed red star in a gold pentagon. By June 29, 1949, 344 officers were commissioned at the Academy,[10] including the teachers of the Faculty of Political Economy, headed by Sándor Ványai and founded in August.[11] This was all the more important as all academies found it hard to recruit teaching staff with up-to-date knowledge, and this had proved most problematic of all at the Petőfi Academy, due to the type and aims of its training. In the Academy's first year, there were three-week courses, three-month and six-month courses, and for two companies (100 students each), ten months' training. After some reorganization, 500 students in five companies began their studies in August 1949.[12]

On August 1, 1949, the Petőfi National Defense Academy for Political Officers' Training became the Petőfi National Defense Institute for Political Officers' Training,[13] known generally as the Petőfi Institute. Instruction in political economy was introduced between August 1949 and April 1950. Further organizational changes occurred on October 1, 1949, when a staff commander, supplies commander and training companies were placed under the Institute commander. The former directed the Training Department and Administrative Section, while the supplies commander headed the Supplies Sub-department. The political officer was in charge of the library and of the Political, Cultural and Sports Department.[14]

Colonel István Otta, the Institute commander,[15] signed on January 1, 1950, a report on the personnel situation,[16] recording that there were 35 political, 28 branch and 13 other officers (totaling 76), 31 subalterns and 66 men currently serving in the Institute. On the student roll were 17 regular officers, 4 regular and 3 nonregular (7) subalterns, and 416 privates, 440 altogether. There were 99 civilian employees. The total strength was therefore 722. On February 3, 1950, the State Security Authority (ÁVH) called for accurate numbers for the Petőfi Institute, in connection with the merger of the ÁVH and the Border Guard.[17] These show that 20 ÁVH men and 13 border guards were studying at the Institute at the beginning of 1950.

On February 4, the Institute submitted draft Organizational Regulations to the Political Chief Section.[18] These stated that the task of the Petőfi Institute was to provide, under Political Chief Section direction, "training for political officers and political leaders of our People's Army, our State Security Authority, and our State Police Force." The Petőfi Institute in general and the areas of personnel and political and military training were subordinate to the Political Chief Section, and its supplies apparatus was subordinate to the supplies commander of the First District National Defense Force Command. The draft structure was as follows:

- The inner command would comprise the Party secretary and the Personnel and D (Security) Departments;

- Directly subordinate to the Institute commander would be the staff commander, Political Department head, supplies commander, and battalion commanders;

- Under the staff commander would be the Training Department, the T (Secret) Administrative Department, the Auxiliary Office, and the Ranks Detachment;

- The head of the Political Department would run the political faculties, the library and the Cultural and Physical Education Sub-departments;

- Under the supplies commander would be the Administration Bureau, Financial Service, catering, building maintenance, uniform/equipment, firearms/ammunition, and transport squads, and the medical section;

- Finally, the battalion commander would control the sub-departments subordinate to him at all times.

The draft gave the Institute commander directing, decision-making and publication rights in service, and political and military training matters. To the staff commander there fell the tasks of a barracks commander. The order on admission guidelines would be the province of the chief of the Political Chief Section, and the consequent procedure would allow future students to be chosen from unit and political officers, political leaders, subalterns, corporals and privates with previous military training in the People's Army, and civilians without military training but with the requisite political knowledge. The draft syllabus of political training was to be issued by the Political Chief Section and the military training syllabus by the commander of the Infantry Branch. These syllabuses would then be processed by the Political and Training departments. The principles of military and political training and education, the ideological and political methodology instructions, the principles for organizing political work, and the length of training would be regulated and published by the Political Chief Section. The same procedural system would apply to extension training of officers and subalterns.

Two methods of passing out were recommended:

- Dismissal (forcible resignation), for deficiencies in general discipline or in progress in political and military studies, or previous life unworthy of a member of the People's Army, or other disqualifying cause;

- Regular release (officer's commission) on successfully completing the political and military studies, with the rank and at a time set by the Political Chief Section.

Political officers would be appointed by the Political Chief Section and military training officers by the National Defense Infantry Branch commander, bearing the demands of the former in mind. The training and education staff could also be augmented by civilian specialist teachers, mainly in the fields of foreign languages and cultural and general knowledge. It was proposed that the right to inspect the Petőfi Institute be held by the minister of defense and the Political Chief Section, and, at their behest or with their permission, by the Infantry Branch commander, in supplies by the Supplies Chief Section or the Budapest First National Defense District Command and the Supplies Chief Section jointly.

On February 21, 1950, Colonel István Otta submitted to the minister of defense a "Draft Program of Party Political Workers' Training Courses of the Hungarian People's Army,"[19] recommending that the course be extended until April 16, to ensure that the envisaged lectures were delivered. It also emerges from the submission that 70 hours of tuition were given to such subjects as "Lenin and Stalin on the organizational bases of the Bolshevik Party," "The Hungarian Workers' Party—Driving Force of the Hungarian People's Republic" and the "Statutes of the Hungarian Workers' Party," and 20 hours to the "Structures and Tasks of Party Organizations."

The teaching of Party political work was so new that even the instructors of the Petőfi Institute were unprepared and requested lecturers to be supplied for the following subjects: the political apparatus of a regiment, battalion and company, and the duties of political officers (8 hours); the working style and method of the Party political apparatus in the Hungarian People's Army (2 hours); Party committees: organs of collective control (2 hours); Party direction of youth organizations in the army (2 hours); keeping Party records in political departments and Party organizations (2 hours); the work of Party organizations in ensuring a vanguard role for Party members in military and political training (6 hours); political activities: the basic form of political education of ranks, officers and subalterns (8 hours); the press: the main means of educating the soldiers of the People's Army (2 hours); Party political work with corporals (2 hours); Party political work to provide the internal and guard services (2 hours); Party political work in attack and defense in the light of the experience of the Soviet Army in the Great Patriotic War (8 hours); lectures by leading staff of the Hungarian Workers' Party Central Committee and leaders of the Political Chief Section on current tasks in political work among the troops (6 hours); the school of Bolshevik education (2 hours); Party political work aimed at ensuring target practice (4 hours); tasks of Communist education of the Hungarian people and the soldiers of the army (2 hours); finally, cultural information work in the People's Army (2 hours). The instructors thought that they were capable of delivering the following lectures if they received a syllabus and possibly new written materials: work to do with membership recruitment for Party organizations (2 hours); ideological/political education of Party members (2 hours); work with Party activists (2 hours); political education of the officer corps of the People's Army (2 hours); political briefings and group and individual discussions (2 hours); lectures, reports, meetings and spontaneous meetings (2 hours); the work of commanders, political departments and Party organizations in strengthening military discipline and political morale (2 hours). The hours of tuition covered lessons or lectures, or a lecture plus a seminar. Interestingly, a distinction was drawn between daytime and evening lectures.

The application to lengthen the course mentioned earlier was probably authorized via a different route, as the Institute commander put forward on April 5, 1950, the plan for the inauguration ceremony for political officers, to be held on April 16.[20] The foot battalion of honor was to consist of a company each from the Police Officers' School, the ÁVH, the Artillery Officers' School, the Signals Officers' School, the Air Force and the Budapest National Defense Guard Battalion. The minister of defense was expected to be guest of honor. After that, and the playing of the National Anthem, the Institute commander would give a short farewell address, and the banner would be brought to the center, where those being initiated would take the oath. Then Lieutenant General Sándor Nógrádi would greet those taking the oath, to which the oldest of those being initiated would respond. The playing of the Internationale would be followed by a parade march and then a "buffet," which the invited guests and just 25 of the inaugurated new political officers would attend.

Again several of the new political officers were posted to the Political Economy Faculty, since the student roll for the next course doubled to ten companies, as one officer who accompanied the banner recalled.[21] In addition, 20 persons were transferred from the Political Chief Section to the Petőfi Institute with effect from April 15.[22]

Colonel Otta described the Institute's mission in a commemorative article in the in-house newspaper, *Petőfi Szellemében*, on May 1:[23]

We soldiers must play our part in the struggle for peace by enhancing the strike capability of our army. The students of our Institute will celebrate May 1 worthily if they undertake to study Soviet combat and Marxist-Leninist political knowledge with the same revolutionary fervor, fortitude and diligence as that with which the workers of the imperialist countries stand heroically in the camp of peace, as that with which our working people build our socialist home country with enthusiasm and determination, led by our great Party and our wise leader, Mátyás Rákosi, and as that with which the soldiers of the Soviet army fought with heroism and resilience in the Great Patriotic War, and as that with which the peoples of the Soviet Union build Communism with enthusiasm, under the leadership of Stalin, the great and wise leader of the world peace camp.

The Institute commander's May situation report[24] reveals that 396 men had been initiated as political officers on April 16, 1950, 6 as captains, 43 as first lieutenants (1 remaining in that rank,) 165 lieutenants, 181 sub-lieutenants, and 1 a chief sergeant. From April 20 onwards, 1,737 persons in groups of 160–280 reported for cadre training, of whom 940 were from the National Defense Force and 138 from civilian life. So there was a shortfall of 22 from the intended roll, solved by May 6, by calling up 40 people who had been conditionally accepted. Again it was indicated that the Institute intended to make up the shortage of political officers by picking out the most politically developed of those completing the course, and it was proposed to the Political Chief Section that these should immediately be commissioned as officers. But it is also clear from the report that "life" was placing incredibly hard tasks

Major General István Otta
(1909–1972)
Inventory number: 32910

before talented young people who still lacked experience of leadership. For instance, "Although the present head of the Political Economy Faculty, Comrade Lieutenant Endre Rataj, is a very talented comrade-in-arms capable of development, he will not become competent to perform the tasks of head of the Political Economy Faculty for another 1–2 years."[25]

The final examinations in military subjects showed that the students were prepared for the duties of a squad commander. But the Institute Command thought the examination board for political subjects too liberal.[26] In any case, the process was "mass-produced," with 16 boards examining almost 400 students in eight subjects over two days. The Institute commander also reported that wide differences of rank and posting among the graduating students had not caused tensions, partly because much attention had been given to this in Party political work, including individual talks with each officer.

"Each of the comrades-in-arms joining the new course shows delight, even effusive delight, at being a student of the Institute. Most have been concerned to conceal any problems or illnesses during medical examinations," was the "triumphant report," although this was gainsaid to some extent by the comment that 11 students had applied for demobilization or return to their units on family grounds.

One of the biggest preparatory tasks for the new course, according to Colonel Otta, was the cadre screening, which involved 11 boards of three or four members screening 1,660 people, of whom 1,190 were found satisfactory, including 200 for "D" (intelligence) work. It emerged that some units had arbitrarily selected candidates to study at the Institute. Some had arrived "protesting vehemently against the idea of being a political officer..." The opinion of one or two boards about some rejected comrades-in-arms was that units "may have sent us their worst..."

The *Petőfi Szellemében*, as an in-house newspaper, was concerned mainly with military, disciplinary, study-related and methodological vigilance, and accommodation issues and proposals for cultural topics. The May 1, 1950, issue, for example, carried a leader entitled "On Discipline," arguing essentially as follows:

> We have to eliminate above all the decisive mistakes that comrades-in-arms in positions of command commit, out of a desire for popularity or out of cowardice, by tolerating neglect of duty or disciplinary breaches of the internal regulations, in internal service or in fulfillment of commands. This leads, in a section or squad where such a comrade-in-arms is the commanding officer, to undisciplined comrades-in-arms constantly urging indiscipline on others and destroying the *good collective spirit*... It will be your prime duty to create and maintain iron discipline in your unit. This can be attained by *example* and by means of education... It is necessary to understand that our *discipline* makes us new-type soldiers, that our discipline is already diametrically opposite to the discipline of a capitalist army.

A student also commented on this:

> Unfortunately, not a day goes by without some breach of discipline, great or small... Comrades-in-arms still can't understand what great importance discipline has in our army; they

don't take the orders given seriously or carry them out without enthusiasm… Let everybody sense that an order, even from a section commander, is given by the working people; it is their command. Let everybody sense their communist duty to carry out the order to the full, for the Party and Comrade Rákosi. The working people stand guard at their workbench or on their land in an honorable and disciplined way. For us to stand our ground well we must learn here at the school—and on leaving here, teach the young people coming into the Army—to be disciplined and obey orders to the full. This now is one of our decisive tasks.[27]

There is plenty of exhortation in the May 30 number as well. A leading article entitled "With the People through Fire and Water" told readers to do the following:

Prepare, learn, and set it as a task before you to be the most disciplined person, a paragon, so that you may fulfill the tasks awaiting you in the spirit of your oath. Obey the order of the Comrade Minister to create iron discipline in your unit, so that you may carry out the instructions of our Party as a true communist, so that you may make our People's Army capable of high-level iron discipline and training, so that at the side of the Soviet Union it may defend through fire and water the other people's democracies and the Hungarian people engaged in building socialism, in the struggle against the imperialists.[28]

Attempts were also made to use the family and workplace to encourage students to study more diligently. *Petőfi Szellemében* often published letters such as this: "My dear son… Study as you studied and worked here at home, that we may be sure to overcome the imperialist camp. Power is in our hands, my dear son, and we will not release any of it, even if it costs us our lives. We must defend our country at all costs, for this country is ours and we are building it for ourselves... We all send you kisses, Jancsika too, but especially your mother. Freedom!"[29] Nor was the commentary lacking, of course. It ran, in part, "The mother, the Hungarian mother speaks to us. She is anxious for her son, but a mother who will protect her free homeland by tooth and nail, and if need be, with the sacrifice of her son, sent this admonitory warning. A message has been sent by the Hungarian mothers, who follow in the footsteps of the parents of Zoyas and Oleg Koshevoys."[30]

The Communist Youth League (DISZ) was launched on June 16–18, 1950, also within the National Defense Force,[31] and soon appeared at the Petőfi Institute.

On June 30, the leader in *Petőfi Szellemében* dealt with the topical issue of success. A quotation of Stalin was placed under the article: "The Red Army won because it had in its ranks the commissars, those tireless organizers and agitators, whose work bound the ranks of Red Army soldiers together like cement, injecting into them the spirit of discipline and heroic valor."[32] The same issue offered assistance in "furnishing the rooms for cultural information work,"[33] suggesting that the following subjects be presented:

- The aims and achievements of the Soviet Union's five-year plan;

- The aims and achievements of the Hungarian People's Republic's five-year plan;

- The importance of the victory of the Soviet Armed Forces;

- Achievements in foreign and domestic policy;

- Life of the sub-unit.

The editors added another Stalin quotation to convey what the essence of a political officer should be: "A regimental commissar is political and moral leader of the regiment and prime protector of its material and spiritual interests—if the commander of the regiment is its head, the commissar must be its father and spirit."[34]

The political officers' report for June[35] devoted a large amount of space to the question of morale, especially the observation that wives, mothers or brides of some students were wearing a crucifix or something similar. Talks with students had "brought out very well the fact that a political officer mainly has to begin the struggle against clerical influence at home. It was put [to them] that it is the duty of all to ensure that wives do not send the children to religious instruction…" Then came a letter-writing campaign to contribute to the struggle against clerical reaction and the clergy's influence. The report also reveals that "denunciation of the Sólyom gang had a good influence on officers and subalterns; nor did it affect badly the older field officers."[36] Military training in the month was at single-combat and section level. The greatest defect was inexperience among squad and company commanders and political officers. Another problem was students studying mechanically, by rote, without understanding the material. Although some deterioration in discipline was reported—164 revealed breaches in June as against 142 in May—the disciplinary situation was seen as being much better, as the control and calling to account had increased in that month compared with the "acclimatization" period after induction. Prominent in the misbehavior were poor obedience to commands, negligence with equipment, and breaches of internal regulations; one case involved theft.

Nor was it simple to leave the Petőfi Institute in those days. The following emerges from a decision of the Communist Party Disciplinary Board for teaching institutions:

Comrade Antal Varga wanted a discharge from the Institute… He claimed that he was not managing the course because he was nervous… he did not feel a sense of responsibility for this vocation… he was not prepared to sacrifice family happiness for anybody's interests… he noted that he had no self-aware-ness… he admitted to being an alcoholic. The Disciplinary Board established that Comrade Antal Varga had committed a grave offense against the great cause of the working class [and] that he wanted, in a way not worthy of a communist, to abandon the post where the Party had placed him. He was turned against the Party and by the influence of the class enemy and petit-bourgeois remnants and became a tool of these... The Disciplinary Board also took into consideration the fact that Comrade Antal Varga, if tardily, had been aroused to self-awareness...

And on the same page of the paper was a statement from Varga, the student, as well:

I wanted to be discharged... because I did not appreciate the great sacrifice with which our Party and working class maintain our Institute, so that we may study peacefully and be good communist officers. Stumbling in the dark, I kept individual interests in mind. The Party explained what my life had been in the past... My father, who visited me at the Institute... said 'Son, I can't come here to study, I can't be one of the Institute's political officers of the working class... I'm proud that my son is studying in the officers' institute named after Petőfi and that he will be an officer...' My eyes were opened at last after my talk with my father, and on my way home, I paced with pride beside my father in the Institute uniform, for the struggle for truth had won within me and I will not sway, and I do not want to be discharged, but to hold my place wherever the Party and the working class have placed me.[37]

According to the July political officer's report,[38] morale was good and the student body understood and supported the Central Committee resolution against clerical reaction. The students sensed and understood the importance of the Korean War. As one put it, "The Koreans fight for peace against imperialism on the front and we do the same on the parade ground." Morale in the detachment of rank and file was judged to be much worse, where "there is a certain rotten core of 15–20 comrades-in-arms, from which various destructive comments, clerical statements, and so on, emanate. There is no doubt that not all the 15–20 comrades-in-arms are the authors of such rotten news and subjects of conversation, but we must put our finger upon the elements initiating the destruction. Here, in the light of the fact that many scions of the kulaks, soldiers of suspect Swabian origin, and other elements with a dubious past enter this detachment of rank and file, we will propose that 15–20 comrades-in-arms be transferred." The military training was still at the section stage. Some shortcomings emerged in the practice of defensive combat in the section, and the student commanders were still often indecisive. But there were unacceptable mishaps at a higher level as well. "There were also cases where an exercise was set up in the wrong direction; for instance, Captain Mikolcsó, head of the Combat Faculty, arranged it so that the enemy came from the east." The political training had covered socialist work emulation, the teaching of Marxism-Leninism on the peasant question, topical issues in the production-cooperative movement, planned management of the national economy, immediate tasks in the development of the national economy, and current problems in the Communist Party's peasant policy.

Two of the proposals put forward are revealing, as the head of the Political Department reported:

Since the new phonograph records allocated to the Institute during the month include several examples of cosmopolitan hits, boogie-woogie, samba, swing and other [genres] whose discarding I have proposed, I recommend that all the phonograph records in the National Defense Force be centrally examined and the inappropriate ones discarded...

... For the cadre selection work for the following course, I propose that we avoid inviting non-party members. On the one hand, in my view, it belittles the Party, Party work, and the institution of political officers to want to turn non-party members into leading, responsible Party functionaries in 6–7 months. That goal, in my view, is not realistic. On the other, it has emerged that there is confusion and there are various troubles with a good many of them, so that a significant proportion of them have had to be dismissed from the Institute.

On August 17, 1950, the commander of the Petőfi Institute submitted to the minister of defense a preliminary proposal for reorganization of the Institute in 1950–1951.[39] The basis was to increase the roll to 1,500, divided into two courses. One would have a roll of 1,000 students (two battalions, nine companies) and be stationed at the Petőfi Institute for ten months. The other (500 students in one battalion of four companies) would last five months and be stationed outside the Institute at the György Kilián Barracks on the corner of the Nagy körút and Üllői út, as the 3rd Battalion of the Petőfi Institute. The complement would have to be increased by 27 field and 26 political officers (the latter to be provided out of the present students), 21 subalterns, 32 men and 46 civil employees.

Colonel István Otta, the Institute commander, and Lieutenant Colonel Vilmos Liszt, the Institute political officer, gave a final report on October 1, 1950, on the peace-loan subscription.[40] 2,304,700 forints altogether had been subscribed at the Petőfi Institute, giving a per capita average of 1,503 forints. (Average monthly pay was around 1,000 forints.) Only 13 persons failed to subscribe, while reactionary views were suspected in only one case.

Preparing for the passing-out parade and referring to the minister of defense's order, the Institute commander put forward for urgent promotion 36 officers: 3 captains, 2 first lieutenants, 22 lieutenants, and 9 sub-lieutenants.[41] On the same day, Colonel Otta submitted to his superiors the plan for the banner-exchange and officers' inauguration parade of October 21, 1950.[42] Mátyás Rákosi, who had donated the earlier group banner, would now present the Petőfi Institute with the new red silk banner from the Iron and Steel Works, embroidered with the new coat-of-arms of the People's Republic. The guest of honor was again to be Minister of Defense Mihály Farkas, with a guest list similar to the one in April.[43]

A leading article entitled "Facing New Tasks," by Lieutenant Colonel Vilmos Liszt, the political officer of the Institute, was designed to prepare the new political officers for service after their inauguration.[44] The examinations might be over, Liszt said, but the real big examination would be to apply in practical work the knowledge that had been acquired: "Theory learned is worth as much as can be applied of it. Our Party and our people have placed serious trust in us and expect us to do our duty." "What were these duties?" asked the author, and found several:

- To educate the soldiers in implacable hatred of the enemy and boundless devotion to the people;

- To prevent imperialist spies and saboteurs from penetrating into the units;

- Political and cultural education work must always start out from full and accurate observance of the military force of unity, military discipline, the moral state, military service and the military oath and regulations;

- To run Party branches in a spirit that raises Party members and candidate members in a spirit of Bolshevik idealism and political self-awareness;

- To ensure the directing role for the Party in the DISZ organization;

- To deal with the requests, needs and political mood of the soldiers;

- To pay a great deal of attention to the recruits;

- To study the rich experience of Party political work in the Soviet army, then apply it in the activity of political departments, political officers and Party and DISZ organizations;

- To become real commanders, since political officers must develop militarily and politically even better than commanders, for then only can they become paragons for the unit.

According to the political officer's report for October,[45] morale was good, as the results of the council elections also showed. Students prepared for their end-of-year examinations with great verve, and it was noted with delight that excellent results were obtained. However, "there were some comrades-in-arms who received subaltern ranks and a couple displayed bad morale. The company comrade political officers and the comrade head of the Political Department dealt with the comrades-in-arms and explained to them that their inauguration as subalterns did not mean they could not become officers... This the comrades-in-arms understood, and they undertook to work honorably in their posts so that they might earn promotion to officer rank." The most important event in combat training was a march exercise of about 80 kilometers during a 1st Battalion field exercise.

On October 16, 1950, the Chief Section of Personnel sent a paper to the Petőfi Institute with a register of the civilians being enlisted.[46] These were to be called upon if the quota of students were not filled by applications from the forces. The list contained details of 1,078 people. On October 21–23, the Political Chief Section sent successive transfer orders for the newly inaugurated officers of the Petőfi Institute.[47] By October 23, 653 political officers or subalterns had been appointed mainly to squads, companies, batteries and battalions as political officers, political leaders, Party or cultural instructors, agitators, club leaders, DISZ secretaries, divisional staff political officers, and so on. Forty-two of the new political officers were posted to the Institute itself. According to a document of October 31, 1950,[48] 150 of the political officers passing out of the Petőfi Institute were being posted to the State Security Authority (ÁVH).

On October 25, Colonel Otta issued syllabuses for the one-year and six-month training courses for 1951–1952.[49] He noted in his introduction that three daily hours of private study would be too little; there would be an average of two political lectures a day, each involving 40–60 pages of reading from the classics of Marxism-Leninism. He had not included political economy in the six-month courses, as the intention was to cover aspects of it within the lectures on Marxism-Leninism. The one-year course would run for the ten months from November 15, 1950, to September 15, 1951. Working days would contain seven hours of study led by a teacher and three hours of private study; on Saturdays and before public holidays, there would be six hours of teaching. That meant 1,198 hours of "hard" study and 748 of private study, the former divided into 762 hours of political subjects (63.5 percent) and 436 hours of military (36.5 percent—a subject breakdown below). The history was to be divided into two large subjects: the modern period and the contemporary period. The 46 hours assigned for the first would cover the French Revolution and Napoleonic Wars, Germany in 1815–1848, the French, German and Hungarian revolutions of 1848–1849, the work of Marx and Engels and the First International, the Franco-Prussian War and the Paris Commune, French, German and British colonization in 1870–1914, the history of China, Japan and the United States in 1870 to 1912–1914, development of the labor movement up to 1914, the Balkan Wars, the work of the Second International, and the events of the First World War.

The main contemporary fields studied (64 hours) would be the Great October Socialist Revolution, the labor movement in the capitalist countries in 1918–1923, the Hungarian revolutions of 1918–1919 and the foundation of the Hungarian Communist Party, the COMINTERN in 1919–1923, Germany, France, Britain, the United States, Japan and China in 1924–1929 and 1929–1939, Poland, Czechoslovakia and the Balkan states in 1918–1940, the COMINTERN in 1924–1943, the struggles of the Hungarian Party

The Subject Structure of the One-Year Training Course

Ser. no.	Name of the subject	Total no. of lessons	Out of the total no. of lessons		Note
			Lecture	Practice lesson and group activity	
	I. Political course				
1.	Principles of Marxism-Leninism	240	160	80	
2.	Political economics	120	84	36	
3.	History	110	86	24	
4.	Party policy in the Hungarian People's Army	102	66	36	
5.	International situation after the Great October Socialist Revolution	30	30	-	
6.	Economic and political geography	80	58	22	
7.	Russian language	80	-	80	
	Total:	**762**	**484**	**278**	**63.5%**
	II. Military course				
1.	Tactics	120	16	104	
2.	Topography	50	-	50	
3.	The use of firearms	80	-	80	
4.	Technical training	20	6	14	
5.	Artillery knowledge	16	6	10	
6.	Armored knowledge	16	6	10	
7.	Signal knowledge	16	6	10	
8.	The use of lethal gas and chemicals	10	-	10	
9.	Basic training	32	-	32	
10.	Sports	50	-	50	
11.	Regulations of military service	14	-	14	
12.	Military management	12	-	12	
	Total:	**436**	**40**	**396**	**36.5%**
	Grand total:	**1198**	**524**	**674**	**100%**

of Communists against the fascist Horthy regime, the history of Hungarian people's democracy, World War II and the Soviet Union's Great Patriotic War, power relations since 1945, the labor movement, Communist Parties, and anti-imperialist national liberation movements since World War II, the struggle of the Soviet Union and the people's democracies for peace, democracy and socialism, and Anglo-American fomenters of a new world war.

Economic and political geography would cover the following main topics:

- The world political map before and after World War I, and during and since World War II (10 hours);

- The area, natural conditions, population, national economy, industry, transport system and agriculture of the Soviet Union (16 hours);

- The same for the Hungarian People's Republic (10 hours);

- The same for the people's democracies: Czechoslovakia, Poland, Romania, Bulgaria, Albania, China, Korea and the German Democratic Republic (18 hours);

- The area, population, classes, political parties, form of state, economy and armed forces of capitalist countries: the United States, the British Empire, France, Italy, Spain, Sweden, Finland, Turkey, Belgium, the Netherlands, Switzerland, Portugal, Norway, Denmark, Iraq, Afghanistan, Japan and the South American countries (26 hours).

One day of study a week would be devoted to natural scientific and cultural lectures. The former (18 hours) would cover the stars, the development of the Earth, the structure of matter, peaceful use of nuclear energy, the origin and development of life and the descent of man (Darwinism). The cultural topics (22 hours) would be the bankruptcy of bourgeois culture, views of culture in the classics of Marxism-Leninism, the main questions of principle in Socialist Realism, and the last hundred years of Hungarian literature.

The Subject Structure of the Six-Month Training Course

Ser. no.	Name of the topic	Total no. of lessons	Lecture	Practice lesson and group activity	Note
	I. Sociological study circle				
1.	Principles of Marxism-Leninism	160	102	58	
2.	History	70	52	18	
3.	Party policy in the Army	100	70	30	
4.	Economic and political geography	60	50	10	
5.	Russian language	40	-	40	
	Total:	**430**	**274**	**156**	
	II. Military department				
1.	Tactical instruction (collaboration of branches of military services)	70	-	70	
2.	Topography	40	-	40	
3.	Musketry training	40	-	40	
4.	Formal training	20	-	20	
5.	Sports	24	-	24	
6.	Regulations of military service of the Hungarian People's Army	10	-	10	
	Total:	**204**	**-**	**204**	
	Grand total:	**634**	**274**	**360**	

Budapest, October 24, 1950.
(Colonel István Otta,
Commander of the Institution)

The six-month training course was to involve 167 training days between October 15, 1950, and April 30, 1951, with 7 hours of teaching and 3 of private study on weekdays (6 and 2 hours on Saturdays and pre-public holiday days) on political (430 hours, 67.8 percent) and military (204 hours, 32.2 percent) subjects.

On November 22, the Institute commander submitted draft training plans for the 1950–1951 academic year to the Political

Chief Section and Section for Military Training.[50] The training goals were defined as follows:

- One-year infantry course: training to perfection for squad (where necessary, company) command, to allow perfect leadership of a squad in battle and behind the lines, and such leadership of a company where necessary;

- One-year branch course: the same as above for units equivalent to an infantry squad or company;

- Five-and-a-half-month infantry course: training to perfection for squad commander of an infantry squad in battle or behind the lines.

The November 30, 1950, report on the strength of the Petőfi Institute[51] shows an actual roll of 1 general,[52] 210 officers, 120 subalterns, 188 privates, 1,566 students, and 181 civilian employees, totaling 2,266 persons and 70 horses, which was short of the regular complement by 1 general, 11 officers, 24 subalterns, 34 students, and 21 civilian employees, but over strength by 47 soldiers. The student body consisted of 103 on the one-year course (including 21 women), 79 on the second year of the two-year course (18), 80 on the first year of the two-year course (16), and 31 lecturers (5), making a total of 293 persons (60).

On February 23, 1951, the Petőfi Institute put to the minister of defense a plan for an instructors' training course.[53] The five-month course for 150 students was to last from May 10 to October 15, 1951, with 110 students being selected by the Cadre Department of the Political Chief Section out of existing lecturers, political officers or persons in leading political posts, and 40 being recommended by the Petőfi Institute itself. Most of the lectures would be delivered by the "Comrade Soviet Advisers,"[54] and so a request was made for 6 interpreters conversant with translating lectures in Marxism-Leninism, political economy and history.

The table shows the syllabus structure of the instructors' training course by faculty, in hours (lectures + seminars + private study). Bases of Marxism-Leninism, history and political economy were to be final examination subjects in all classes except the Party Political Work Faculty, where they would be replaced by the Party political work.

Subjects	Marxism-Leninism	Political Economy	History	Party Political Work
Bases of Marxism-Leninism	300 (120+100+80)	160 (96+64+0)	160 (96+64+0)	160 (96+64+0)
Political Economy	100 (72+28+0)	220 (100+60+60)	100 (72+28+0)	100 (72+28+0)
History	100 (64+36+0)	70 (50+20+0)	220 (100+60+60)	70 (50+20+0)
Party Political Work	50 (50+0+0)	50 (50+0+0)	50 (50+0+0)	220 (100+60+60)
International Situation/Conditions	30 (30+0+0)	30 (30+0+0)	50 (30+20+0)	30 (30+0+0)
Economic/ Political Geography	0	50 (34+16+0)	0	0
Bases of Methodology	20 (20+0+0)	20 (20+0+0)	20 (20+0+0)	20 (20+0+0)
Total	600 (356+164+80)	600 (380+160+60)	600 (368+172+60)	600 (368+172+60)

The Petőfi Institute's reorganization proposal for 1951,[55] completed on April 2, increased the load to three courses with a total of 1,630 students. The one-year course was organized into two battalions totaling 1,060 students. The 400 students of the political extension training course were to form the 3rd Battalion.

Major General Otta published a revised plan for the new courses on April 18.[56] The proportion of study time devoted to political subjects was increased by 2 percentage points to 87 percent. The subject structure of the instructors' training course, in hours (lectures + seminars + private study), by faculty, was as follows:

Subjects	Marxism-Leninism	Political Economy	History	Party Political Work
Bases of Marxism-Leninism	304	180	180	180
	(160+94+50)	(120+60+0)	(120+60+0)	(120+60+0)
Political Economy	100	204	100	100
	(72+28+0)	(110+62+32)	(72+28+0)	(72+28+0)
History	96	96	212	96
	(70+26+0)	(70+26+0)	(106+66+40)	(70+26+0)
Party Political Work	60	60	60	184
	(50+10+0)	(50+10+0)	(50+10+0)	(86+72+26)
International Situation/Conditions	30	30	38	30
	(30+0+0)	(30+0+0)	(30+8+0)	(30+0+0)
Economic/Political Geography	0	20	0	0
		(20+0+0)		
Bases of Methodology	10	10	10	10
	(10+0+0)	(10+0+0)	(10+0+0)	(10+0+0)
Total	600	600	600	600
	(392+158+50)	(410+158+32)	(388+172+40)	(388+186+26)

The plan for the inauguration ceremony for political officers on April 28, 1951,[57] contained several novel features. The ceremony would be in several stages. First would come a ceremonial issue of the order from 13:30 to 15:00, where the minister of defense's promotions order would be read and the "School Commander" personally hand new shoulder marks to the new captains, and then, as the order was read out further, each battalion commander and political officer would distribute shoulder marks to first lieutenants, lieutenants, and sub-lieutenants, and possibly subalterns.

There was some tension over the introduction of new, Soviet-style uniforms. Some were pleased, but opinions were also heard saying that they "would sooner have stayed with the old one than have the new. Or… one student remarked that the workers would never agree to the soldiers walking about in Russki uniforms. The comrades' case is still under investigation; the Party Organization is dealing with them. In the 7th Company, a case of listening to the Voice of America radio station occurred during the month, and the comrade officer assigned to the company was among those listening; this has been investigated by the Political Department, and the Institute Assembly will deal with it."

There was satisfaction with the results of the end-of-year examinations for the six-month training course, with averages of 3.86 for the bases of Marxism-Leninism, 4.07 for history, 3.89 for political economy, and 4.09 for Party political work. The distribution of grades among the classes was as follows:

Grade	Bases of Marxism-Leninism	Political Economy	History	Sport
5	140	143	175	261
4	151	181	171	165
3	169	145	114	45
2	18	10	6	0

Grade	Unit Training	Party Political Work	Combat	Topography
5	239	148	216	58
4	203	205	205	43
3	42	114	67	18
2	0	9	1	0

11 persons who received grades of 2 in political subjects were ordered by Lieutenant General Nógrádi to remain at the Institute for a further month and repeat the material.[58]

On May 25, the Petőfi Institute prepared a 29-page "Information Memorandum" on the activity of the institution so far.[59] The introductory historical review recalled that before the foundation of the Petőfi National Defense Institute for Political Officers' Training in March 1949, political officers had been trained in the Petőfi National Defense Academy for Education

Officers.[60] "Up to August 1949, the Education Officers' Academy and the Petőfi Academy had been marked by an almost total absence of Party activity, crudeness of teaching methods, a low level of political and military training, uncomradely behavior, and a high level of indiscipline. Most of the field and political officers stationed here before August 1949 have since left the army and many have been brought before the courts for hostile activity."[61] The authors asserted that the institution had undergone steady development since August 1949, and by November 15, 1950, it was providing two five-month courses, for 500 and 1,100 students. The material for political education was the subject matter used at the Central Party School. For the next five-month course (500 students) and one-year course (1,100 students), launched on November 15, 1950, the Petőfi Institute turned, on the orders of the minister of defense, to the system of the Soviet institutes for political officer training, with the aid of Soviet advisers. In presenting the education work, the memorandum chose to mention the subject of "Petőfi, political officer of his time" as one of the lectures designed to nurture the traditions of the War of Independence and progressive traditions.

Although mention has already been made of the courses in the 1950–1951 training year, this memorandum provides new, accurate data on the subject. The following types of training were being given at the Petőfi Institute in May 1951:

- A one-year course for political officers. This had begun on November 15, 1950, with 1,100 students, of whom 53 had dropped out by the last third of May.

- A five-month course for instructors of political officers. This had begun on May 15, 1951, with 154 students in 4 sections: Marxism-Leninism (78 students), political economy (25), history (28) and Party political work (23).

- An extension training course for political officers with 230 students.

The document also divulges statistical information on the students of the one-year political officers' course according to several criteria, for instance the following:

A. Origin: 39.2 percent workers (23.1 percent workers in large factories); 48.1 percent peasants (10 percent cooperative members; 3.6 percent poor peasants; 32.2 percent new peasants; 2.3 percent middle-ranking peasants); 3.6 percent petit bourgeois; 0.5 percent intelligentsia; 8.6 per cent miscellaneous;

B. Occupation: 65.2 percent workers (14.3 percent workers in large factories); 23.1 percent peasants (4.2 percent cooperative members; 0.4 percent new peasants; 18.5 percent poor peasants); 0.5 percent intelligentsia; 2.2 percent petit bourgeois; 9 percent other;

C. Party membership: 63.3 percent members; 36.6 percent candidates; 5 persons non-party;

D. Date of membership/candidacy: 1945: 152 persons; 1946: 109; 1947: 192; 1948: 147; 1949: 126; 1950: 315;

E. Educational attainment: 4 or fewer years of elementary: 2.6 percent; 5–6 years of elementary or 1–2 of middle school: 45.8 percent; 7–8 years of elementary or 3–4 of middle school: 49.3 percent; high school leaving certificate: 2.3 percent;

F. Age: 17: 0.3 percent; 18: 5.5 percent; 19: 16.8 percent; 20: 24.6 percent; 21: 22.4 percent; 22–25: 27 percent; 26–30: 3.4 percent.

These figures also show that political reliability was the main criterion for admission in the autumn of 1950, relegating general education to the background. So 15–20 percent of the course students had very great difficulty with the material, especially at the beginning of the course, and with understanding the political subjects. This was one reason why 4–5 percent of the students were discharged.

There were 240 officers at the Petőfi Institute at that time, including 44 political officers, 87 political lecturers, and 109 field officers. All the political officers were Party members, but only 76 percent of the field officers were Party members or candidate members. The vast majority of the officers had served at the Institute for less than a year; almost without exception, they were very young and inexperienced, with hardly an officers' school or a five-month Petőfi Institute course behind them. On May 1, 1951, a further 10 field and 43 political officers arrived.

On July 4, 1951, Lieutenant General Sándor Nógrádi, the chief of the Political Chief Section, issued an invitation to apply for the Petőfi Political Officers' Academy,[62] arguing that the development of the Hungarian People's Army required political officers to become real masters at Party political work. The two-year Political Officers' Academy Course starting in the fall of 1951 to prepare officers for regimental political officers' posts and above. He was therefore calling for political officers to apply for this. Candidates had to be at least 24 years old and to have completed at least the eight grades of primary education, as well as a Petőfi Institute course, and have at least one year's experience as a political officer, although the last requirement could be waived by the chief of the Political Chief Section. The invitation named as subjects Marxism-Leninism (20 questions based on the intermediate cadre-training course), the Hungarian language (spelling, expression), arithmetic (the four basic calculations with integers and decimal numbers), geography and the Service Regulations.

On August 1, 1951, Major General István Otta reported on a military course at the Party College,[63] attended by students of the one-year and two-year Party College courses and by teaching staff there. The students had been placed in a battalion of three companies, including volunteer "female comrades-in-arms" and teachers, the latter training alongside the seminary students. The instructors were officers of the Petőfi Institute (a battalion commander and political officer with company and squad commanders). Company commanders taught combat, and squad commanders taught the other subjects (firearms, topography, and so on). Held in January–June 1951, the course had comprised 140 hours of military training in four weekly theoretical lessons at the Party College and a monthly eight-hour exercise—tactical deployment on four occasions and shooting on two. There were final examinations on tactics, topography and firearms, with the three companies achieving averages of 4.74–4.82, and the college teachers an average of 4.95. On July 21, the minister of defense promoted 20 students to major, 41 to captain, 63 to first lieutenant, 45 to lieutenant, and 7 to sub-lieutenant.

The July monthly report[64] stated that the company averages had risen by 0.1–0.4 to 4.1–4.35. Morale was rated as being good, although two students had been expelled from the Party and the Petőfi Institute for demoralizing activity, and there had been other "destructive manifestations; these the comrades-in-arms had noticed and reported to company political officers…" One student, for instance, had asked in conversation with his fellows "whether they knew who I. Michurin [the biologist][65] was. The comrades-in-arms could not say, whereupon he had given the answer: 'Comrade Rákosi's father, because he made a talking squash [slang for penis].'" He of course also underwent Party disciplinary proceedings and was recommended for dismissal from the Institute.

Plans for the political officers' inauguration parade on September 16, 1951,[66] largely conformed with earlier procedures, except for the fact that the Institute commander intended to distribute "excellent student" badges as well as shoulder marks. The guest of honor would be Lieutenant General Sándor Nógrádi, and the Soviet national anthem would be heard alongside the Hungarian national anthem and the Internationale.

Morale was again said to be good in September,[67] as evinced by the examination preparations, peace-loan subscription, preparations for People's Army Day, and assessments of the international situation. The 1st Battalion of the one-year course scored an average of 4.26 in the finals and the 2nd Battalion an average of 4.16, giving an Institute average of 4.21. The political officers on their extension training course took state examinations in Marxism-Leninism and Party political work. In military training, the tactics examination average for the one-year course was 4.19 and the firearms average 4.18. The companies of the extension course attained averages of 4.4 and 4.21 for tactics, and 4.11 and 4.34 for firearms. The peace-loan subscription at the Petőfi Institute had been closed on the evening of October 2, 1951, with an aggregate of 1,237,700 forints being subscribed, as against a target of 953,419 forints.[68]

On October 23, 1951, the Institute commander submitted a proposal to the minister of defense for a Correspondence Faculty,[69] prompted by the need for "officers and generals to obtain higher military-political training while continuing to serve without interruption," as Colonel General Mihály Farkas put it. It would be a three-year course for 75 students freed from duties to attend for one whole day and two evenings a week (16 hours). In addition, there would be study leave of a month at the end of the first and second years and two months at the end of the third. They would be excused all other military and political extension training. The syllabus would consist of three groups of subjects: 1,130 hours of social science subjects covering the bases of Marxism-Leninism (280 hours), dialectic and historical materialism (160 hours), political economy (200 hours), general history (160 hours), Hungary's history (120), Party political work in the Hungarian People's Army (160 hours) and the international situation (50 hours); 870 hours of military subjects, consisting of tactics (300 hours), military topography (80 hours), shooting (120 hours), artillery (40 hours), armored and mechanized units (40 hours), engineering (40 hours), chemical defense (20 hours), signals (30 hours), formal training (50 hours), air force (20 hours), physical education (80 hours), military administration (20 hours) and service regulations (30 hours); 400 hours of general educational subjects, consisting of economic and political geography (100 hours), Russian language (240 hours) and literature (60 hours). Of the 2,400 hours of study, 1,920 would be lectures and 480 private study or preparing for examinations (29 altogether over the three years). This subject matter matched the syllabus and program of the All-branch Faculty. The students would be officers and generals who were Communist Party members, with secondary school educational attainment or "approaching that," and experience in Party politics and military command. The course would start on November 15, with an examination period of August 1–31 and leave to be taken between September 1 and November 1.

The October political officers' report[70] states that the new students began to arrive at the Institute on October 8. Of those with grades of "not satisfactory," 30 were offered a transfer to the subalterns' school, which they accepted. Of the 1,355 people applying to the Institute, 848 were accepted and another 15 were still under consideration.

Following the January 17, 1951, Communist Party Central Committee resolution mentioned earlier, the minister of defense issued on November 1 the order founding the Stalin Political Academy. This looks more like a cheap political brochure than the founding document of an institute of higher education. The letter from Mátyás Rákosi greeting the new Academy was in a similar style, although there were some more sober sentences, such as the following:

> We have founded the Stalin Political Officers' Academy alongside the Military Academy of the People's Army to allow commanders and political officers of the National Defense Force to increase continually, alongside their military knowledge, their political knowledge and general education, and meet requirements for the enhanced tasks that face Communist commanders and trainers with the development of our army.[71]

O R D E R
No. 055
OF THE DEFENSE MINISTER
OF THE HUNGARIAN PEOPLE'S REPUBLIC.
Budapest, November 1, 1951

Comrades!

Based on the resolution of the Central Committee of the Hungarian Workers' Party and the Government of the Hungarian People's Republic, I order the foundation of the Political Officers' Academy. In line with the resolution of the Party and Government we shall name the Political Officers' Academy after Comrade Stalin, the inspired pursuer and developer of the cause of Lenin, the organizer of the Soviet Armed Forces, the founder of the brilliant Soviet military science, the true friend and helper of the Hungarian people.

So shall the Academy bear the proud name the "Stalin Political Officers' Academy."

Our people, true to the teachings of Lenin and Stalin, are successfully building social-ism in our country under the guidance of the Hungarian Workers' Party and our beloved leader Comrade Rákosi. Application of the teachings the great Stalin shows that our people's army is becoming an ever stronger preserver of the freedom, independence, and peace of the Hungarian people; that the whole people's army is imbued with a spirit of proletarian inter-nationalism and ardent devotion to our country.

Bearing the name of the great Stalin signifies very great honor and responsibility. The students of the Academy must become like our paragons, the heroic commissars brought up by Lenin and Stalin, of whom Comrade Stalin said:

"The Red Army won because it had in its ranks such exceptional organizers and agi-tators as the military political commissars, who welded the lines of Red soldiers together with their work and took discipline and fighting morale into their ranks."

The task awaiting the command of the Stalin Political Officers' Academy is to raise political officers who set an example of sacrifice, discipline, communist willpower, and high-level intellectual political education of the troops, and can undertake to the full the com-mands of our Party and people in defense of the country.

May the Stalin Political Officers' Academy bear its great name proudly and raise high the triumphant banner of Marxism-Leninism!

I order:

that the corps of commanders, political officers, and teachers of the Stalin Political Officers' Academy so prepare for the tasks facing the Academy that the students train themselves mil-itarily and politically, consciously, striving for ever more perfect study results, knowing no exhaustion, and carry out the study plan to the full;

that they unceasingly reinforce and deepen in the complement of the Academy unswerving loyalty and devoted adherence to the working people, our great Party and its beloved leader Comrade Rákosi, the liberating Soviet Union, and the great Stalin;

that they raise from the students of the Academy political officers who are brave and capa-ble of every sacrifice for the cause of the people and the international working class and of execution of the commands;

that the students of the Stalin Political Officers' Academy, while developing without pause their political and military professional knowledge, put all their strength into raising the level of their general education;

that they be living followers and preservers of all the progressive national and military traditions of combat, for which our worthy ancestors have spilled their blood and offered so many victims down the centuries;

that they go more deeply into studying the teachings of Marxism-Leninism, the policy of our Party and the Stalinist military traditions, and become unshrinking, disciplined, brave warriors and qualified leaders of the masses of soldiers.

Praise to the great STALIN, the great friend of the Hungarian people, the gifted leader of the peace camp!
Long live the organizer of the victorious socialist building, the trainer and stimulator of our People's Army: the Hungarian Workers' Party!
Long live the wise teacher and beloved leader of the Hungarian people: MÁTYÁS RÁKOSI!

MIHÁLY FARKAS, lieutenant general (ret.),
defense minister of the Hungarian People's Republic

One Petőfi Institute student analyzes in a short article the political conditions in the period being examined, under the title "We in Our Battle Positions Struggle against the Titoite Rats."[72] Similar sentiments appeared in short articles by students of the Stalin Academy entitled "We Respond with Study to the Blatant Provocations of Tito's Band" and "We Can Have No Task so Urgent as to Make Us Forget Vigilance."[73]

Major General István Otta, the Academy commander, turned to his superiors on November 21, 1951, with a request that the Stalin Academy should receive a banner from the workers of the Stalin Ironworks, to mark Stalin's birthday on December 21.[74]

According to the November report of the Stalin Academy,[75] morale there and in the Petőfi Institute was good. Those concerned had shown great responsibility in ensuring that the academic year would begin smoothly and in screening the new students and assigning them to companies. Teaching began as planned on November 1 and study was being performed with great enthusiasm, although some of the Academy students especially had serious problems on account of being separated from their families, as their housing problems had not been solved. Some were even requesting a return to their units. The Party branches had begun to deal with these. Some of the Institute students were more alarmed by the difficulties of study. There were some who had requested a transfer from the six-month to the one-year course, saying that they could not master the material in six months. One student who ran away from the Institute was sentenced to five years' imprisonment by the military court. There was some tension among the officers of the 3rd battalion billeted in the Kilián Barracks, due to the pressure of work and the number of duty periods caused by the officer shortage.

The banner from the workers of the Stalin Ironworks at Sztálinváros (now Dunaújváros) was received by the complement of the Stalin Academy on December 21, 1951. The ceremony was followed by a banquet to mark Stalin's 72nd birthday.[76]

Some thoughts were quoted earlier from the letter of a mother to her son. The official correspondence of wives cannot have been too easy to write either, judging by this example.[77]

Dear Lali,

I received your card and I'm rushing to answer it. You ask in your letter what the news is from home. The greatest you know: I mean the speech of Comrade Rákosi. We all heard it too in the factory. I thought of you immediately. You too will certainly be glad about the end to the rationing system. It's freed us of many minutes full of much care. On Monday, after the speech, I went among my colleagues with such pride. This was a victory for communists, for the Party. We and the comrades thought gladly of our father, Comrade Rákosi, who looks after us with zealous care. We talk about you soldiers too. We don't doubt that our weapons are in good hands; we have nothing to fear in the future…

Your loving wife, Magdi.

Yet, as "class warfare intensified further," student relations acquired a strange ambiguity. The in-house paper was full of articles on comradeship in a military and a communist sense, on helping others, while other pieces gave reminders about the "Responsibility to keep military secrets unfailingly."[78] The same great secrecy showed in the way in which authors ceased to append a rank to their articles or the ranks of those mentioned, just the name and the attribute "officer" or "senior officer."

On January 16, 1952, came a visit to the Stalin Academy by Mátyás Rákosi.[79] "The few hours that our beloved leader spent amongst us strengthened us," according to a leading article in *Petőfi Szellemében*. "They strengthened us because we know that Comrade Rákosi always goes where people are doing especially important work for our people and our motherland. We are preparing to defend our Academy, our motherland and our people; we are preparing to be educators and inspirers of our People's Army. The visit by our paramount commander and dear teacher, Comrade Rákosi, has underlined the weight of this great honorable task."[80] The paper saw it as a major mission to play an active and consistent part in tightening discipline, but meanwhile it continued to scourge itself or exaggerate the role of Party organizations. One example was headlined "Party Leadership of T Group to Blame for Latenesses,"[81] in an article that began by saying "We have emphasized day after day that our Institute timetable is an inviolable law…" Then it describes anonymously how "the comrade student X. Y., disobeying the order from our people, was 2 hours 15 minutes late. Did he consider… when he was late, how he had committed a grave crime against our people? … But the students mentioned are not solely responsible for these faults. They are clearly responsible primarily. But the leadership and membership of the [Party] branch are responsible as well." One student wrote a long article in the same issue entitled "The Deputy Commander Can Only Gain Respect from Subordinates by Ignoring No Single Case of Indiscipline."[82]

The commander of the Stalin Academy submitted to his superiors on February 4 a proposal for relocating the Petőfi Institute,[83] in which the Party control over the army is quite apparent: "According to the January 23 resolution of the MDP Secretariat, 'the Petőfi Political Officers' Training Institute must be divided off from the Stalin Academy in the fall of this year and reorganized as a separate two-year institute for training political officers.'"

The February 16, 1952, issue of the in-house paper devoted several pages to a Party meeting at the Academy on the previous Sunday. On the subject of lecturers, the report stated that although the faculties could meet the demands made of them, the problem was that lectures and group classroom activities were "not instructive enough. There is a damaging abstraction to be found in some lectures." The report went on to name the lecturers who not only lectured, but gave regular help to students in their studies, with positive effects, of course. Major General Otta, addressing the Party meeting, noted as the worst fault the fact that "the study is still mechanical. Some of the students aim only for a good grade and do not learn in the meantime. We have to train our students to be people who boldly ask about what they do not understand and vigilantly watch over the purity of theory." The poor shooting results were explained by saying "Not all our political officers shoot well either. If the students see that the political officers shoot well, they too will do everything to become good shots."

One student tried to urge his comrades on to better performance in an article entitled "On Liberation Day, We Have to Show in Our Parade, in a Manner Worthy of the Soviet Warriors, the Strength to Deter the Imperialists!"[84]

The exorbitant search for enemies led the Academy commander to recommend three persons for dismissal on March 17, 1952.[85] All three had World War II experience; two were highly qualified teachers of tactics; the third had completed the Higher Commander Course at the National Defense Academy and was currently deputy commander of the Stalin Academy. The groundlessness of the proposal is clear in the last case, for instance, from his subsequent record: he reached the rank of major general and a deputy minister's post, becoming one of the decisive leaders in the Hungarian People's Army.

Before the inauguration of April 1952, *Petőfi Szellemében* sought to give parting advice to those graduating from the Petőfi Institute. For instance, the paper called on the new political workers "never to forget for a moment that a political officer is the Party's representative, representing the Party in the units of our people's army. Through him the Party educates our warriors in ardent patriotism, implacable hatred of the imperialists and their lackeys, heroic resistance, courage, understanding love of our weapons, honor and honesty."[86] The new political officers were also reminded to demand discipline, and not to tolerate slackness or indiscipline, while never forgetting that "officers and men with us are united in spirit, education and origin." And yet another piece of advice: "increasing demands can only be met by those who train themselves further, in a planned, systematic fashion."[87] The in-house paper also retained on its agenda the question of raising the standard of study in the Stalin Academy.[88]

The Education Council of the Petőfi Institute likewise examined the experiences of the first half-year, and based on these, recommended briefly "Let us study more independently and more deeply."[89] Students at the Stalin Academy recommended establishing "Russian circles" among students, to help them to learn the Russian language more rapidly and successful. This was followed soon afterwards by a contest between them.[90]

Again in the early days of June 1952, political officers were inaugurated at the Petőfi Institute whose "day-to-day task will be to educate the sons of our working people and thereby strengthen our army."[91] One officer warned the students of the Institute that "the road to good study results led through daily reading of the *Szabad*

Nép [the Communist daily paper],"[92] arguing that this was "not newspaper reading as such, but study… Those who study the paper attentively day after day can follow domestic and foreign policy events, see the connections between things, and see the likely direction in the development of events as well." This line of thought was taken up by an Academy student: neglect of newspaper reading would lead to falling behind events, becoming ill-informed and unable to take a correct position, and so on.[93]

In July 1952, the in-house paper lost the name *Petőfi Szellemében*, borne for five years, becoming instead *Sztálin Útján* (On Stalin's Road), "as a consequence of the vast development undergone since the foundation of the young Stalin Academy, which abounds in valuable achievements, and with which the Academy's paper has closely to catch up."[94] In the "Take It as an Example" column on page 2 of the same number was a picture of the Academy student Ferenc Kárpáti,[95] "taken before the unit banner for his excellent results. Academy Student Kárpáti's head has not been turned by praise. His present work also shows that he has earned the distinction. While doing his Party work well as Party secretary, he stands his ground excellently in study work as well. He is exemplarily disciplined. He keeps to the revision plan point by point. He regularly revises for twelve hours a week…"

On August 1, 1952, came a major structural change: the Petőfi National Defense Institute for Political Officers split from the Stalin Political Officers' Academy and each went its own way. From November 1, 1952, the Petőfi Institute would provide basic political officers' training, and what was now to be the Stalin Military Political Academy offered higher training for political officers.[96] In the same month, the staff of the two institutions congratulated Mátyás Rákosi in a telegram on his election "by the most ardent desire of the whole Hungarian people as speaker of our National Assembly and president of the Council of Ministers of our People's Republic… We inform Comrade Rákosi on this festive occasion that we have completely fulfilled our pledge made on the occasion of his 60th birthday. We promise, in a way worthy of your sons, to tighten our discipline, enhance our battle readiness, achieve an average above 4 in our end-of-year examinations, and double the number of excellent and vanguard students by comparison with the results at the end of the half-year."[97] The end-of-year examinations duly brought group study averages of 4.07–4.48. The N Group, for instance, had a sixfold increase in the number of excellent students.[98]

Yet it became increasingly clear in the fall of 1952 that such rapidly increasing expectations could not be met in two years. So the length of the Stalin Academy Course (and the National Defense Academy Course) was lengthened to three years in the 1952–1953 academic year.[99] Interestingly, the Stalin Academy in-house paper paid no regard to this important qualitative change, preferring to carry a telegram from Rákosi on the opening of the renamed institution, underlining the need for political as well as military knowledge:

> We have established the Stalin Political Officers' Academy alongside the Military Academy of the People's Army to allow the commanders and political officers of the national defense force to develop continually, alongside their military expertise, their political knowledge and their general education, so that they can meet the enhanced requirements facing communist commanders and instructors with the development of our army. Bearing the name of the great Stalin binds the commanders, political officers and students of the Academy… to utilize Marxist-Leninist theory—learned at the Academy in a spirit of internationalism combined with true patriotism—in practical life, in resolving the tasks of combat training, and if need be, in battle as well.[100]

Greater care seems to have been taken with launching the academic year and establishing the requirements at the Petőfi Institute. *Petőfi Szellemében* (entering its sixth year) now returned to being the paper of the Petőfi Institute, as it became separate on October 31, 1952. It carried a leading article entitled "Before the New Training Year,"[101] summarizing the main expectations of the staff and lecturers:

- All officers and subalterns in any position whatever had to require of students at all times impeccable dress, cleanliness, punctuality, honesty and responsibility;

- They had to refrain from empty rhetoric and from being unprepared;

- They had to acquaint themselves fully with students' characteristics and capabilities;

- They had to be capable of arousing and sustaining interest in the subject taught;

- Theoretical instruction had to be tied to practice to a maximum extent;

- They had to strive constantly to expand their methodological skills as lecturers;

- Training should proceed from simple to complex, from known to unknown, and from details to whole;

- They had to underline the development of communist morality in themselves and their students, impart to their students socialist patriotism, socialist humanism, a spirit of community, strong will, and strength of character, and develop in them a sense of beauty, taste, interest in the arts, a love of music, and a love of literature and drama.

Observing these, in the author's view, would raise the Institute's political, moral and professional standards.

The Petőfi National Defense Institute for Political Officers' Training, often now referred to as the Petőfi Political Officers'

School,[102] launched one-year and two-year courses in the 1952–1953 training year.[103] There were 225 teaching days (45 for private study) in both the first and second years of the two-year course (November 1, 1952–October 1, 1953 and November 1, 1953–September 15, 1954), giving 1,440 teaching hours (810 of private study) in each. There were to be eight hours of lectures and two

The Subject Structures of the One-Year and Two-Year Petőfi Institute Courses in 1952–1953/4, Hours per Course

Subject	Two-Year Course	One-Year Course
I. Social Sciences		
Bases of Marxism-Leninism	500	300
Political Economy	280	200
Global History	240	-
Hungarian History	144	-
History	-	88
History of the Hungarian Workers' Party	92	72
Party Political Work in the Hungarian People's Army	210	120
International Situation	50	36
Total	*1,516*	*816*
II. Military Subjects		
Tactics	240	120
Topography	80	54
Armaments and Shooting	160	100
Artillery Studies	40	24
Armored/Tank Studies	34	20
Engineering Studies	40	24
Signals Studies	36	22
Aircraft Studies	20	-
Chemical Defense	20	10
Formal Exercises	86	40
Physical Education	80	40
Health Training	10	-
Military Administration	18	-
Service Regulations	40	20
Total	*904*	*474*
III. General Education		
Economic and Political Geography	100	60
Russian Language	220	80
Hungarian Language and Literature	80	-
Arithmetic	60	-
Total	*460*	*140*
Grand Total	**2,880**	**1,430**[104]

hours of private study on weekdays. There were ten hours also on Thursday, the day for private study, and eight hours on Saturday. Social science subjects took up 53 percent of the time, military subjects 31 per cent, and general education 17 percent. (See the table on the previous page.) The one-year course ran from November 1, 1952, to September 15, 1953, with 179 days of study (44 of private study), giving a total of 1,432 hours of study (980 of private study). The subject structure was much the same as for the two-year course: 57, 33 and 10 percent respectively.

Petőfi Szellemében published a little article by one of the language teachers, entitled "Learning the Russian Language Is a Fine Task,"[105] arguing that knowledge of the Russian language would tie them to the freedom-loving peoples of the Soviet Union and the peace camp, and give greater knowledge of the militant patriotism of the Soviet people and its firm, steadfast commitment to Communist victory.

During the process of "disaggregating" the minister of defense's 1952–1953 training order to institute levels, meetings of activists were held on November 27, 1952. The first such meeting at the Stalin Academy opened with references to the resolutions of the Nineteenth Congress of the Communist Party of the Soviet Union, especially the need to enhance criticism and self-criticism, and wage an implacable struggle against errors. It was recommended that the Academy's execution of the tasks in the minister's order be examined in that light. The speaker warned everyone against imagining that the imperialist danger had lessened. Far from it, for West Germany and Japan were rearming, while the nearby "Tito and Ranković gang"[106] were totally subservient to the Americans. The equivalent meeting at the Petőfi Political Officers' School[107] was specific about the tasks deriving from the minister's order, although here too they were based upon the Soviet Nineteenth Congress resolutions. Speakers specified what General Mihály Farkas had instructed the officers' schools to do: completely

eliminate irregular occurrences and tighten discipline, so as to supply the Hungarian People's Army with officers possessing ever-
greater knowledge and practical experience.[108]

On December 10, 1952, the Central Committee dealt with
political work in the Hungarian People's Army, in the light of shortcomings. The need was for comprehensive extension training for
the army's political apparatus, stricter requirements of political
officers, and an increase in the length of the Petőfi Political
Officers' School Course to three years. By the 1953–1954 academic year, only candidates with a high school certificate would be
admitted to the Petőfi.[109] However, the social structure was seen as
favorable; 73 percent of the political officers in 1952 came from a
worker background, 13 percent from peasant families, 11.1 percent
from the families of white-collar employees, and only 2.9 percent
from those of officials or artisans.[110] The real problem seemed to be
with the level of education. Bearing these requirements in mind,
stronger study discipline became a central issue at the Petőfi.[111]
Resolution on raising the level of education prompted the launch of
a correspondence high school course at the Petőfi, but it was soon
found that only 20–30 percent of students attended the consultations: "It is the task of the students of the correspondence high
school and preparatory course to show, by appearing promptly and
well prepared at the next consultation, that implementing the Party
resolution is compulsory for every communist alike."[112]

Based on the government resolution of November 5, 1952, the
minister of defense issued an order on December 27 replacing the
co-command system with one of one-man, individual command,
with effect from January 20, 1953.[113] The former co-commander
political officers became political deputy commanders. Both institutions naturally paid close attention to this fundamental change in
the status, activity and influence of the political deputy commanders. On January 23, 1953, *Petőfi Szellemében* published a whole
page of extracts compiled by Major General Jenő Hazai, deputy

head of the Political Chief Section, entitled "Lenin and Stalin on One-Man Leadership." He quoted Lenin as saying "How can we ensure the strictest unity of purpose? By subordinating the will of thousands to the will of one man." According to Stalin, "Ensuring one-man leadership is an effective weapon... against petit-bourgeois laxity and indiscipline, in the struggle for the good organization of the proletarians and for discipline." Although Rákosi was not mentioned in the title, he was quoted too: "Strict one-man leadership, iron discipline and high-level organization are decisively important in the army. One-man leadership is one of the most important factors in the structure of our armed forces." From all this, Hazai drew the conclusion that one of the main tasks of the political bodies was to help the commander by every means to ensure high-level results in combat and political training, military discipline and organization, and further reinforcement of order.

The new tasks and methods deriving from the introduction of individual command was the subject of an activists' meeting at the Stalin Academy on January 21, 1953. Again it was stated that the political organizations had a duty to give maximum support to the many-sided activity of the commander and to teach the subordinates obedience to the individual commander. The commander, of course, was likewise required to give full support to the political officers and organizations.[114] For obvious reasons, the subject remained on the in-house paper's agenda for some time to come.

A government resolution of August 8 granted college status (that of an educational institution offering a diploma course of higher education) to the Stalin Academy and the National Defense Academy.

Stalin's death on March 5, 1953, brought immediate changes for the Soviet Union and its captive nations in Eastern and Central Europe. The decisive turn for Hungary came at an "inter-Party meeting" to which the Soviet Bolshevik leadership summoned a delegation of Party and state leaders. Two features of its constitution pointed towards the trend. Mihály Farkas was omitted, and Imre Nagy, hitherto somewhat in the background, was included. Rákosi was sharply criticized; fundamental changes were ordered; Imre Nagy was to take over as prime minister. Preparatory work in the highest echelons of the Hungarian Workers' Party, especially the resolution at a Central Committee meeting on June 27–28, brought condemnation of grave errors committed in recent years and acceptance of a "June Program" of reforms. Among the economic aspects censured were the forced pace of industrialization, especially in heavy industry, where the country's natural resources situation had been ignored, the neglect of agricultural production, the excessive speed of agricultural collectivization, the neglect of consumer-goods production, and thus the decline in the standard of living. It had been a grave mistake to wind up the firms of artisans and small traders; some villages were left with no barber. Arbitrary, violent measures had been taken and protests at them ignored. Collective leadership had given way to personal rule. Few "of Hungarian descent" had been tolerated in the leadership, power being seized by a clique of "Muscovites": Rákosi, Ernő Gerő, Farkas and József Révai. [115] Often Rákosi had personally ordered acts of terror by the secret police (ÁVH). The resolution called for the following: reducing the scale of industrialization and restoring its internal balance, so that heavy industrial development was no longer pursued at the expense of light industry and consumer goods production; enhancing development, profits and production in agriculture, with support for private farming, a moratorium on the debts of collective farms, and a more humane system of compulsory agricultural delivery quotas; making consumer goods more easily available to all; building over 10,000 dwellings in 1954–1955; cutting the strengths of the armed forces and the swollen state bureaucracy; restoring the legality of the judiciary; allowing internal deportees to return to their homes; legislating for an amnesty for political convicts and the disbandment of the concentration camps.

The new Parliament elected on May 17 met for the first time on July 3–4 and duly elected as prime minister Imre Nagy, whose decided intention it was to implement the June Program or New Course.[116] Detail on Imre Nagy's first government and its policies is necessary here first because if he had been allowed to carry through the June Program, it would have led towards the kind of "communism with a human face" that many great intellectuals in Eastern and Central Europe were envisaging at that time: a social and political system in which the government would ensure human dignity and basic human rights for its citizens. More consistent effort in that direction might have precluded the outbreak of the 1956 Hungarian Revolution. But the Rákosi clique made a comeback in 1955, removing Nagy from office and expelling him from the Communist Party, and breaking off the New Course, all with Moscow's consent.

After the announcement of the June Program, "signs of discontent and outrage at the crimes and mistakes committed emerged with increasing force in the Academy as well. As the wrangling, lack of principle and crimes in the Party leadership came to light, so did the suppressed criticism from below. Strong condemnations were heard in various Party forums, in a way hitherto unheard of at the Academy, initially over the unbearably bureaucratic, tense atmosphere created by some leading bodies in the institution (the Personnel Department, for instance), extending later to the maximalist study requirements, which most of the students were unable to meet.[117]

The examinations of the three courses at the Stalin Academy ended the academic year in September 1953. Most of the students attained grades of "Good" and "Excellent," although some sections had averages (3.54 and 3.61) much weaker than the overall average (3.95 and 4.04).[118] On September 30, the state examinations began, providing for the first time a chance for successful students to obtain a college-level diploma. A leader in the in-house paper

intimated to officers soon to be holding new, responsible posts that
they should do the following:

Retain forever in your memories the gravity of the errors to
which the absence of collective leadership and un-Party-like
chieftainship could lead. You should secure freedom of criticism with all your might. Watch conscientiously over the
observance of the rights of warriors. Thus and only thus can
you ensure that the Communists and the whole complement
are bound together in the great work of training, and that the
Party organization is truly the leading force.[119]

A special extension raining course for teachers was introduced
at the Stalin Academy in the 1953–1954 academic year.[120] There
was still an acute shortage of thoroughly trained military teachers,
even though the overall strength of the army had been greatly
reduced, especially in the fall of 1953.[121]

The Third Congress of the Hungarian Workers' Party in May
1954 was also enlisted to step up the study activity. Students at the
Petőfi successfully carried out their tactical training before returning for their final and state examinations in July. Here it was mentioned again that work and study discipline and thorough, planned
study needed tightening further.[122] Despite the great impediment of
taking part in the flood protection required in the summer of 1954,
there could be no reduction in the requirements: "If the staff of our
school understand the importance of the examinations and our students study in full knowledge of their responsibility to the Party,
then success, victory, will again be ours, as it has been so often in
this training year."[123]

The Political Chief Section of the Ministry of Defense and the
Academy Command, intent on modernizing the teaching, decided
in the summer of 1954 to draw also on the experience of those who
had completed the two-year course at the Stalin Academy. It was
stated during the exchange of views that the knowledge gained at

the Academy had proved a great help in practical work. The high standard of military expertise had helped to consolidate the reputation of political deputies. But it was also intimated that a greater proportion of the teaching should go on general educational subjects.[124] Some prestige for the Stalin Academy was gained from an article by the Kossuth laureate Academician István Rusznyák, president of the Hungarian Academy of Sciences, writing in the in-house paper: "The Stalin Academy will soon be issuing fresh graduates… aware of what the new intelligentsia means to a Hungary building socialism… The training is now of an incomparably higher standard in an army whose officers are men with a broad outlook, many-sided training and a love for science and the arts." He added that "we workers in science" were certain that "the directing and educating work of the officers graduating from the Academy will be reflected in a rapid rise in the standard of our army."[125] As recognition of the higher standard of professional and political attainment, an order of April 26, 1955, from the minister of defense regularized the issue of Academy badges to graduates of the Stalin and Miklós Zrínyi Academies. These were duly presented on October 14, 1955, to the first officers to have completed the three-year course,[126] although a shadow may have been cast by anxiety about forced retirement, as a government decision on September 1 to reduce the strength of the Hungarian People's Army by another 20,000[127] began to be implemented. Development of the institutes of higher military education was promoted again by a government decision[128] that 50 percent of the fall 1955 officer intake would have to have completed high school or the equivalent.

On September 5, 1956, came an order from the minister of defense amalgamating the Stalin Military Political Academy and the Petőfi Political Officers' School, as the Sándor Petőfi Military Political Academy. A few days previously (on September 1) Major General István Otta had been consigned to the reserves and appointed by the Communist Party Central Committee as general

Major General Endre Pesti
Inventory number: 91525

secretary of the Hungarian Soviet Society. Colonel Endre Pesti was appointed as commander of the Sándor Petőfi Military Political Academy.[129]

Shortly after that appointment, events developed into a revolution and war for independence that scholars in Hungary and abroad have compared in significance to 1848. Hungary's foremost poet, György Faludy, wrote a poem on it entitled "1956, You Star." It tested the loyalty of all persons and institutions to the nation, the students and staff of the Sándor Petőfi Military Political Academy being no exception. It showed the futility of intensive Bolshevik indoctrination, since many of its students and faculty members joined in their compatriots' struggles, although all too many followed the instructions of the old regime and fought against their own people.

Understanding the significance of what happened in the Sándor Petőfi Military Political Academy calls for a short summary of the events of 1956. The Hungarians were not seeking a revolution in 1956, but the vast majority of them sought fundamental changes. The university students drafted a "16-Point Program of Youth" that included in its demands national independence, a democratic bill of rights for citizens, the elimination of Communist terror, a review of political trials, rehabilitations and the return of prisoners of war still held in the Soviet Union, the prosecution of Rákosi and Farkas, restoration of national symbols and holidays, the return of Imre Nagy to government, the removal of Stalinists, elimination of Hungary's colonial status, and non-intervention in Hungary's domestic affairs. But the program did not demand the abolition of the Communist Party, whose organization and policies it sought to democratize. Nor was there a demand to abolish socialism. Subsequent massive student-instigated demonstrations were met with armed resistance by the regime on October 23, when it was demanded that the 16 Points be broadcast on Radio Kossuth. The youth met violence with violence, and the Revolution started. By October 28, the freedom fighters were victorious. Imre Nagy, as the new prime minister, declared an armistice and started implementing the 16 Points. No person, party or significant group wished or dared to challenge the establishment of independence and democracy. The Soviet Union started in the morning hours of October 31 to wage a war against Hungary with overwhelming forces, and Hungary was the loser in that first war between socialist states. But the ideas of 1956 could not be forgotten, and became victorious after all in 1989, when Hungary's Third Republic was declared. A democratically elected Parliament took over in 1990 and started to realize the ideas of 1956. In 1999 Hungary joined NATO, and in 2004 the European Union.[130]

Despite pressure from superiors, some teachers at the Sándor Petőfi Military Political Academy began to make contact with the university student organizers as early as October 22, 1956. These contacts may have been a factor behind the initiative of Lieutenant Colonel László Lukács,[131] studies deputy to the Academy commander, and others to call a meeting of the Party leadership of the social science faculties on the morning of October 23, 1956. The faculty Party secretary stated in his introduction "We cannot be idle about the demands of the university youth either, for we too are a college!" It was decided to call an improvised meeting at nine and invite the Party organizations and subalterns as well. As preparation for this, the Party leadership drew up an appeal to the university youth. Although there was also a proposal that it should be made to the Central Committee, not to the university students, and should request clarification of the political situation, this was postponed to a later time, the Party leadership insisting that they should turn to the university students at that point. On the other hand, it was proposed that the appeal should dissociate the Petőfi Academy from what were seen as incorrect demands by the university youth and support only rightful ones. All that the Party leadership would concede, however, was to insert the word "rightful" in the appeal. The essence of this appeal, presented to the emergency meeting that followed, was this: "We the subalterns and students of the military political academy named after the ardent freedom fighter Sándor Petőfi agree with the rightful and fair demands of the Hungarian university youth. We support your rightful and fair demands. We are sons of the people; we have sworn on the people in our oath. We fight beside the people through fire and water." After several other contributions ("nor can we be silent any longer," "we have to show our colors" and "we have to take a stand"), the meeting endorsed the appeal with great enthusiasm. It was decided that it should be delivered to the radio and the press; delegations would march to the universities to make the appeal known, and another appeal would be made to the Central Committee during the morning hours.

The "director responsible in principle for compiling the content of the letter" to the Central Committee "clearly had proposals for the letter's content prepared. Some members of the committee wanted to include in it demands for the removal of several army leaders (Bata, Hazai and Hídvégi). Some members went beyond that to demand that Rákosi... be taken to court immediately, and so on. An attempt to oppose this course was made by [the Academy commander] and a few other comrades, by [whose] influence the demands mentioned and some other exaggerated demands were omitted from the letter, and a relatively moderate letter was compiled." The Petőfi Academy delegations met with great enthusiasm at the Technical University and Military Engineering School, the Agricultural University and the College of Physical Education, to shouts of "The army's with us!" On the other hand, they asked the Academy people to express their support for the youth demands by being present at the afternoon demonstration.

The afternoon assembly of the Petőfi Academy put specific proposals on taking part in the demonstration, arguing that "The presence of Communist officers will provide a guarantee for avoiding provocations." After all that, about 250–300 of the students and subalterns arrived in Bem tér in marching order singing revolutionary songs, where they met with great loud cheers. But soon, increasingly decided demands were being made for removing red flags, red stars and the arms of the People's Republic, which persuaded some of the contingent to return to the Academy, while the others followed the crowd to Parliament.

At 5 p.m. on October 23, 1956, the Academy commander ordered those on the premises to go on alert. Then he received a phone call during the evening from Colonel General István Bata, the minister of defense, asking whether the authorities could rely on the Academy, in the light of the situation that had arisen in the city. The reply was that they could, and not long afterwards, there came an order from the Ministry of Defense to send from the Petőfi Academy 150 men with rifles and bayonets, each with 40 rounds of ammunition, to reinforce the garrison at the Hungarian Radio building, restore order, and ensure that the speech by First Secretary Ernő Gerő was broadcast. The minister of defense gave a clear order to the Academy commander that it was forbidden to shoot at the people, but this bred indecision and chaos among the volunteers for the task. The Academy did not possess so much as an armored car, and there was a delay of over an hour before the contingent could leave the Academy a few minutes before eight o'clock. On the corner of Múzeum körút and Bródy Sándor utca, they ran into a hostile crowd that jostled the commander of the contingent and demanded that it withdraw. The prohibition on using their weapons meant that they were obliged to turn back for the Academy, to applause and cheers from the crowd.

When this incident was reported to the Ministry of Defense and the chief of General Staff, another firm order was issued to carry out the task, but the ban on "shooting at the people" remained. The detachment tried to get round this huge contradiction by taking pistols and tear-gas and hand grenades in their pockets and marching to the Radio building behind the national flag, singing 1848 revolutionary songs. The Ministry of Defense people promised that they would inform those defending the building of this and ensure that they were admitted, but this did not happen. Instead they were met by warning shots, tear-gas grenades and fire hoses. Only 35–40 of them managed to get into the building. The rest returned to the Academy or mingled with the crowd.

Problems and confusion were compounded by the indecision of superiors, who issued "requests" instead of orders. Even the defense of the premises was left to the "discretion" of its commander, Colonel Pesti, who therefore deployed three mortars and a heavy machine gun used for instruction purposes but in working order. Ammunition was drawn independently from the arsenal at Törökbálint.

On October 27, firing broke out between the Academy defenders and an armored car from the Széna tér rebel group. This cost the lives of a civilian army driver and four rebels, with four other rebels wounded. On the same day, a squad from the Academy accompanied the Soviet advisers and their families to safety. Morale was lowered further when a Ministry of Defense order came forbidding the use of weapons even if the post were attacked.

With the victory of the Revolution on October 28, a Provisional Revolutionary Military Council was formed at the

Petőfi Academy to keep the personnel informed, end the Academy's isolation, and ensure cohesion and unity there.

The old Sándor Petőfi Military Political Academy then became the Faculty of Political Work at the Miklós Zrínyi Military Academy. From the spring of 1957 onwards, it continued what the Revolution had interrupted, educating political officers for the army, in a spirit of loyalty to the Soviet Union and the Communist Party, which had taken the name Hungarian Socialist Workers' Party on November 1, 1956.

The Revolution and the behavior of political officers during it were the subject of many investigations. A summary report in the spring of 1957 noted the following: "The personnel of the Sándor Petőfi Military Political Academy, and its faculty members not least, were not unaffected by the wave of revisionism that swept over the country… Under the banner of struggle against mistakes and shortcomings revealed by the Twentieth Congress, members of the faculty, when clarifying some problems in the social sciences, used anti-Marxist, revisionist ideas." The report carries the names of faculty members who had spoken up especially for this trend: five officers in the Department of History of the Soviet Communist Party and four in the Department of Universal History, five in the Department of Philosophy, two in the Department of Party Political Work, and one in the Department of Hungarian History. Prominent among the students were the second year in the Department of Education. It was clear that leading faculty figures there had dominated the cooperation with the revolutionaries. The report concludes that creative methods of solving theoretical problems were replaced by unprincipled disputes and petit-bourgeois chatter. The true and imagined mistakes of the past were exaggerated. Hidden behind the supposed struggle against the cult of personality and dogmatism was a degradation of real achievements and efforts to discredit the leaders of the Party, state and Peoples' Army. Certain persons questioned the leading role of the Ministry of Defense's

Political Chief Department in the armed forces. This report was followed by reprisals against ostensible culprits, who were dismissed from the Academy and the army and expelled from the Communist Party.[132]

These circumstances together show how those responsible for the dependability of the Peoples' Army had failed to learn from the 1956 Hungarian Revolution and War for Independence, although by no means forgetting that it had occurred.

Notes

1. *Magyarország hadtörténete (2)* [Military History of Hungary II], editor-in-chief Ervin Liptai (Budapest, 1985), p. 548.

2. László Csendes and Tibor Gellért, *Kronológia a Honvédség történetéből 1945–1990* [Chronology of the History of the *Honvédség*, 1945–1990] (Budapest, 1993), p. 74. It is interesting to note a comment by Dénes Felkai, in the author's possession, in Sándor Mucs and Ernő Zágoni, eds., *A Magyar Néphadsereg története 1945–1950* [History of the Hungarian People's Army 1945–1950 (*sic*, actually 1959)] (Budapest, 1979), p. 8, of a letter, stating that the "Petőfi" Academy had started operation on September 1, 1948 (actually *de facto* on September 1, but *de jure* on December 12, as has been seen).

3. According to Major General Dr. Ervin Liptai (ret.), "Dénes Felkai was not 'born an officer,' but he was an excellent, absolutely genuine man with a broad education. A large part in his dismissal was played by the connection that he had had with László Rajk, since they had fought together in Spain" (p. 1 of a letter of May 26, 2004, in the author's possession).

4. Information from Colonel Dr. Antal Oroszi (ret.).

5. *Magyarország hadtörténete*, p. 545.

6. Csendes and Gellért, *Kronológia a Honvédség történetéből 1945–1990*, p. 79.

7. *Honvédségi Közlöny* 7 (1949). Researched by Colonel Dr. Antal Oroszi (ret.).

8. This is presented rather differently by Felkai, for on pp. 9–10 of his contribution, with some sorrow, he writes the following: "But who were those political officers who actually worked excellently? The education officers, who had struggled hard since 1945 for education in a socialist spirit in the army, for whom their ties to the coalition now ceased, to whom Comrade Nógrádi referred sarcastically at the official opening of the Petőfi Academy on December 12, 1948, as 'so-called education officers.' So the view could survive to this day that Comrade Nógrádi expressed in his book [*Új történet kezdődött* [A New Story Begun] (Budapest, 1966), pp. 193–194]: 'The task was no small one: to build out of nothing a new people's democratic

army with a strong revolutionary spirit…' In fact the facts quoted above make clearer the crimes of Mihály Farkas and his associates: they urgently rid themselves of Comrades Pálffy, Sólyom, Illy, and so on, and so theirs is seemingly the credit for creating the Hungarian socialist army!"]

9. Mucs and Zágoni, eds., *A Magyar Néphadsereg története 1945–1950*, p. 211.
10. Csendes and Gellért, *Kronológia a Honvédség történetéből 1945–1990*, p. 83.
11. Colonel Endre Rataj, "Politikai gazdaságtan és hadigazdaságtan tanszék" [The Political Economy and War Economy Faculty], *Akadémiai Közlemények. Az akadémia szerveinek története II. kötet Társadalomtudományi ágazat tanszékeinek története.* 120 (1985) 2: 37. Former students also entered the other social science faculties at the same time. See Liptai, op. cit., p. 1.
12. Rataj, "Politikai gazdaságtan," p. 38.
13. Csendes and Gellért, *Kronológia a Honvédség történetéből 1945–1990*, p. 86.
14. *Ibid.*, p. 88.
15. See the Biographies of Key Personalities.
16. Honvédelmi Minisztérium Hadtörténeti Intézet Petőfi Politikai Tisztképző Intézet [Ministry of Defense, Military History Institute, Petőfi Institute for Political Officers' Training] (hereafter cited as HM HL PPTI), 07/I./T. eln. szü.–1950 (Oct. 1).
17. *Ibid.*, Pság 0104–1950 (Jan. 27).
18. *Ibid.*, 0138/Szt. tö.–1950 (Feb. 4).
19. *Ibid.*, 02117. sz.–1951 (Aug. 31). As the dates show, the "mills" of the Political Main Group ground slowly, as the document did not arrive back at the Petőfi Institute until 18 months later.
20. *Ibid.*, 0374/Szt.–1950. (April 5).
21. Rataj, "Politikai gazdaságtan," p. 41.
22. HM HL PPTI 0390/T.Szü.–1950 (April 12).
23. This in-house paper appeared at varying intervals and underwent various "phases of development," as will be seen. At this time it was classed as "for service use only" and carried under the banner the slogan "WITH THE PEOPLE THROUGH FIRE AND WATER."
24. HM HL PPTI 0481/T.Pol.–1950 (May 5).
25. It may be that Lieutenant Endre Rataj was still unprepared at that time to organize a new faculty, but he was to show gratitude for the

positive assessment mentioned earlier by serving for several decades as a colonel, as head of the Miklós Zrínyi Military Academy's Political Economy Faculty, later the Political Economy and War Economy Faculty. He led the process of introducing the subject of War Economy.

26. According to Colonel Dr. Liptai (ret.), "The army was developed at too fast a pace in those years. The number of young men suitable and prepared for an officer's career was far fewer than desirable. So in setting requirements, great concessions were made also in admissions, and more deplorably still, in graduations." Liptai, *op. cit.*, p. 2.

27. *Petőfi Szellemében*, May 20, 1950, p. 3.

28. *Ibid.*, May 30, 1950, p. 1.

29. *Ibid.*, p. 3.

30. *Ibid.*, June 15, 1950, p. 3.

31. Csendes and Gellért, *Kronológia a Honvédség történetéből 1945–1990*, p. 99.

32. *Petőfi Szellemében*, June 30, 1950, p. 1.

33. *Ibid.*, p. 3.

34. *Ibid.*, p. 4.

35. HM HL PPTI 0705/T.Pol.o.–1950 (July 1).

36. Lieutenant General Sólyom, the chief of the General Staff, was executed on false charges. In 1956 he was rehabilitated. "As for the 'older field officers,' the poor fellows felt deeply intimidated and dared not express an honest opinion. It soon emerged that with few exceptions and no individual assessment, they would be brutally treated and dismissed from the army." Liptai, op. cit., p. 2.

37. *Petőfi Szellemében*, July 15, 1950, p. 4.

38. HM HL PPTI 0834/T.pol.–1950 (Aug. 2).

39. *Ibid.*, 0905/Szt.–1950 (Aug. 17).

40. *Ibid.*, 01114/Pol.o.–1950 (Oct. 1).

41. *Ibid.*, 01164/T.Szü.–1950 (Oct 10).

42. *Ibid.*, 01165/Szt.–1950 (Oct. 10).

43. See Note 21.

44. *Petőfi Szellemében*, Oct. 20, 1950, p. 1. Apart from that, the "enhancement of class struggle" referred to on the front page may also be reflected in the fact that this was the second number where the "For internal use only" warning on the banner was followed by "It is forbidden to take this off the Institute premises."

45. HM HL PPTI 0834/T.–1950 (Nov. 3). The October 3 date on the document should read November 3.

46. *Ibid.*, 01214/T.szü.–1950 (Oct. 20).

47. *Ibid.*, 01228; 01229; 01230/T. szü.–1950 (Oct. 21); 01232; 01239; 01240; 01241/T.szü.–1950 (Oct. 23).

48. *Ibid.*, 01325/T.szü.–1950 (Nov. 8).

49. *Ibid.*, 01252/T.–1950 (Oct. 25).

50. *Ibid.*, 01324/kik. o.–1950 (Nov. 22).

51. *Ibid.*, 024016/T.szü.–1950 (Nov. 30).

52. István Otta was promoted to major general on November 7, 1950. László Csendes and Tibor Gellért, *Háborútól a forradalomig 1945–1955. Adalékok a magyar hadsereg történetéből* [From War to Revolution, 1945–1955. Contributions to Hungarian Army History] (Budapest, 1994), p. 96.

53. HM HL PPTI 0552/T.–1951 (Feb. 23).

54. "The first adviser alongside the commander of the Petőfi Academy was Colonel Kostenko, a clever, flexible, well-prepared man enjoying general respect. Later, some social science faculties received advisers. They helped much with the 'professionalism' of faculty work, but were less experienced men of lesser caliber, who found it hard to grasp Hungarian conditions." Liptai, *op. cit.*, p. 3.

55. HM HL PPTI 0773/Szt.–1951 (April 2).

56. *Ibid.*, 0870/T.–1951 (April 18).

57. *Ibid.*, 0840/Szt.–1951 (April 20).

58. "Having passed out, poorly performing students commissioned with unprincipled liberalism undermined respect for political workers and officers, as a burden on the army and society alike." Liptai, *op. cit.*, p. 3.

59. HM HL PPTI unnumbered–1951 (May 25) 79/a.–98/a. f.

60. There is much confusion about the foundation of the institution, in specialist literature and in the recollections of its founding commander.

61. HM HL PPTI unnumbered–1951 (May 25) 79/a.–98/a. f.

62. 02079/HVK.Pol.–1951. Researched by Colonel Dr. Antal Oroszi (ret.).

63. HM HL PPTI 01866/T.–1951 (Aug 1).

64. *Ibid.*, 01898/T.Pol.–1951 (Aug 3).

65. Ivan Vladimirovich Michurin (1855–1935), Soviet botanist.

66. HM HL PPTI 02158/Szt.–1951 (Sept. 3).

67. *Ibid.*, 02403/T.–1951 (Oct. 2).
68. *Ibid.*, 02402/T.–1951 (Oct. 3).
69. *Ibid.*, 2728/Int.Pk.–1951 (Oct. 23).
70. *Ibid.*, 02864/T.–1951 (Nov. 5).
71. *Petőfi Szellemében*, Nov. 7, 1951, p. 1.
72. *Ibid.*, Nov. 17, 1951, p. 15.
73. *Ibid.*, Dec. 1, 1951, p. 1.
74. HM HL PPTI 0936/T.–1951 (Nov. 21).
75. *Ibid.*, 03224/T.–1951 (Dec. 5).
76. *Ibid.*, 03255/T.–1951 (Dec. 12).
77. *Petőfi Szellemében*, Dec. 15, 1951, p. 4.
78. *Ibid.*, Jan. 12, 1952, p. 1.
79. "Rákosi behaved relatively naturally on the visit, but all the more hysterical were the preparations for it and the interpretation put on the events, which were a nauseating example of the cult of personality." Liptai, *op. cit.*, p. 4.
80. *Petőfi Szellemében*, Jan. 19, 1952, p. 1. There is no telling whether it was as an outcome of Rákosi's visit—for he gave liberal advice wherever he went in the Academy—that the banner of the next issue replaced the Petőfi quotation "WITH THE PEOPLE THROUGH FIRE AND WATER" by "FOR OUR PEOPLE, FOR OUR PARTY, FOR OUR BELOVED MOTHERLAND."
81. *Ibid.*, Jan. 26, 1992, p. 5.
82. *Ibid.*, p. 7.
83. HM HL PPTI 0287/T.–1952 (Feb. 4).
84. *Petőfi Szellemében*, March 1, 1952, p. 5.
85. HM HL PPTI 0669/T.Szü.–1952 (March 17).
86. *Petőfi Szellemében*, April 19, 1952, p. 1.
87. HM HL PPTI 01053/Pság–1952 (March 25).
88. *Petőfi Szellemében*, May 1, 1952, p. 2.
89. *Ibid.*, May 10, 1952, p. 1.
90. *Ibid.*, p. 2; *Ibid.*, May 17, 1952, p. 4.
91. *Ibid.*, June 5, 1952, p. 1.
92. *Ibid.*, June 19, 1952, p. 3.
93. *Ibid.*, July 3, 1952, p. 3.
94. *Sztálin Útján*, July 24, 1952, p. 1.
95. Ferenc Kárpáti, later colonel general and minister of defense.
96. Csendes and Gellért, *Kronológia a Honvédség történetéből 1945–1990*, p. 123.

97. *Sztálin Útján*, Aug. 20, 1952, p. 2.

98. *Sztálin Útján*, Sept. 24, 1952, p. 2.

99. Mucs and Zágoni, eds., *A Magyar Néphadsereg története 1945–1950*, p. 244; Dr. Lajos Móricz, "A magasabb tisztképzés magyarországi történetének elvi, szervezeti és vezetési kérdései" [Theoretical, Organizational and Leadership Questions of the Hungarian History of Higher Officer Training], *Akadémiai Közlemények* 178 (1991): 20.

100. *Sztálin Útján*, Nov. 7, 1952, p. 3.

101. *Petőfi Szellemében*, Oct. 31, 1952, p. 1.

102. The names were still used variously, but the new "'Petőfi' Political Officers' School" was gaining ground. See, for instance, *Petőfi Szellemében*, Nov. 30, 1952, p. 1.

103. HM HL Sztálin Pol. Ti. Ak. [Stalin Political Officers' Academy] 01826/1952. Researched by Colonel Dr. Antal Oroszi (ret.).

104. *Sztálin Útján*, Nov. 7, 1952, p. 5.

105. *Petőfi Szellemében*, Nov. 14, 1952, p. 5.

106. *Sztálin Útján*, Nov. 29, 1952, p. 1.

107. *Petőfi Szellemében*, Nov. 30, 1952, p. 1–2.

108. Móricz, "A magasabb tisztképzés magyarországi történetének szervezeti és vezetési kérdései," p. 21.

109. Mucs and Zágoni, eds., *A Magyar Néphadsereg története 1945–1950*, p. 255.

110. *Ibid.*, p. 249.

111. *Petőfi Szellemében*, Jan. 16, 1953, p. 1.

112. *Ibid.*, p. 2.

113. Csendes and Gellért, *Kronológia a Honvédség történetéből 1945–1990*, p. 125.

114. *Sztálin Útján*, Jan. 23, 1953, p. 1.

115. That is, the group of pre-war and wartime exiled Communists in Moscow, not the underground Party at home. The four men were of Jewish descent. Echoes of "anti-Zionist" measures taken in Hungary as in the Soviet Union during Stalin's last months continued as the New Course began to unfold.

116. The *novy kurs* was orchestrated from Moscow, with very similar measures being introduced concurrently in several satellite countries and Soviet republics. For a fuller account and documentation, see Lajos Izsák, "Imre Nagy's First Government, 1953–1955," in Lee Congdon, Béla K. Király and Károly Nagy, eds., *1956: The*

Hungarian Revolution and War for Independence (New York, forthcoming in 2006), Part I, Chapter 2.

117. Liptai, *op. cit.*, p. 4.

118. *Sztálin Útján*, September 30, 1953, p. 2.

119. *Ibid.*, p. 1.

120. Rataj, "Politikai gazdaságtan," p. 44.

121. Csendes and Gellért, *Kronológia a Honvédség történetéből 1945–1990*, p. 136.

122. *Petőfi Szellemében* VII: 15, p. 1. The date July 14, 1954, was left off this issue and the next, July 28, was also numbered VII: 15.

123. *Petőfi Szellemében*, July 28 1954, p. 1.

124. *Sztálin Útján*, July 2, 1954, p. 3.

125. *Sztálin Útján*, Sept. 18, 1954, p. 1.

126. Csendes and Gellért, *Kronológia a Honvédség történetéből 1945–1990*, pp. 148 and 151.

127. Mucs and Zágoni, eds., *A Magyar Néphadsereg története 1945–1950*, p. 275.

128. *A magyar állam szervei 1950–1970. Központi szervek* [Bodies of the Hungarian State, 1950–1970. Central Bodies] (Budapest, 1993), p. 372.

129. Rataj, "Politikai gazdaságtan," p. 43; Csendes and Gellért, *Kronológia a Honvédség történetéből 1945–1990*, pp. 163–164.

130. For more detail, see Béla K. Király, *Wars, Revolutions and Regime Changes in Hungary, 1912–2004. Reminiscences of an Eyewitness*, eds. Piroska Balogh, Andrea T. Kulcsár and Tamás Vitek (New York, 2005), pp. 323–329. See also Károly Nagy, ed., *Information Bulletin No. 2. Significant Documents of the Hungarian Revolution of 1956* (Budapest, 2006).

131. Lukács died defending the Radio building and received a posthumous colonelcy.

132. HM HL PPTI, 1956 Coll. I, Section 386–423 ff.

Chapter X
The Miklós Zrínyi Military Academy in 1955:
Political Events and Military Conditions

The many important international political events in 1955 took an inconsistent course, with concurrent, alternating indications of détente and tension in the political and military measures and counter-measures, gestures and threats. Here are some of the major *positive* influences on the international climate:

- A decree by the Soviet Presidium formally ending the state of war with Germany (January 25);

- Successful Soviet–Austrian negotiations on a state treaty (April 12–15);

- The arrival of a Chinese government delegation under Chou Enlai for the Bandung Conference (April 18), heralding the Non-Aligned Movement of the 1960s;

- The visit to Belgrade by a Soviet Party and government delegation (May 3);

- The signature of the Austrian state treaty, restoring sovereignty and equal state rights but precluding future moves towards political or economic union with Germany (May 15);

- The Soviet cession of the Port Arthur naval base to China (May);

- A Moscow visit by a Yugoslav Party and government delegation headed by President Tito (June);

- Tito's reception of Indian Prime Minister Nehru and President Nasser of Egypt on the island of Brioni (July);

- The Geneva conference of US, British, French and Soviet heads of state on Germany, European security and disarmament, and ties between the two world systems, bringing marked international détente without concrete advances (July 18–23);

- US–Chinese talks at ambassador level in Geneva on bilateral issues (August 1);
- A Soviet decision to reduce its armed strength by 640,000 men (August 31);
- An agreement to establish Soviet–West German diplomatic relations (September 13);
- The departure of the last Soviet occupation troops from Austria (September 19);
- The treaty on relations between East Germany (the GDR) and the Soviet Union (September 20), followed by Yugoslav recognition of East Germany (October 10);
- Austria's admission into the United Nations (October 14) and legislation on Austria's "eternal" neutrality (October 26);
- The Geneva conference of US, British, French and Soviet ministers of foreign affairs (October 27–November 16).

Also frequent were political and military events that *impeded* the détente process:
- The Soviet repudiation of the Soviet–French treaty of alliance (May);
- The Baghdad Pact, creating the CENTO Middle East military bloc (February);
- Britain's accession to the Western European Union (May 6) and retaliatory Soviet repudiation of the 1942 treaty of mutual assistance with Britain and similar agreements with France (May 7);
- Admission of West Germany into NATO (May 9–11); the Soviet Union established the Warsaw Pact with Albania, Bulgaria, Czechoslovakia, East Germany, Hungary, Poland and Romania (May 14) one day before the signature of the Austrian state treaty; under the latter, the Soviet Union's right to move troops into Austria via a corridor through Hungary ceased, but Soviet troops remained stationed in Hungary under the provisions of the Warsaw Pact;

- West Germany broke off diplomatic relations with Yugoslavia after the latter recognized the East German regime (October 19);

- A Soviet–Polish agreement on the legal status of Soviet troops "temporarily" stationed on Polish soil (December 17).

These cardinal international events naturally affected the *domestic* political climate in Hungary as well. For instance, a Hungarian Party delegation (of Mátyás Rákosi, Imre Nagy, Mihály Farkas, Lajos Ács and Béla Szalai)[1] was summoned to Moscow on January 8 by the Presidium of the Communist Party of the Soviet Union, where General Secretary Khrushchev stated that "Imre Nagy represents 'anti-Party' views and the CPSU president takes a position of support for Rákosi's group."[2] This stance encouraged the Hungarian Party Central Committee on March 2–4 to conclude that the correct decisions of June 1953 had been "distorted by certain people in an opportunistic, anti-Marxist way, leading to harmful right-wing errors and right-wing deviation."[3] Imre Nagy was blamed for this, and fabricated accusations were used by the Central Committee on April 14 to strip him of all his functions. He was voted out of office as prime minister four days later. He was succeeded by András Hegedűs, with Ernő Gerő as deputy prime minister.[4] However, the Central Committee also struck out leftwards on April 14, excluding Minister of Defense Mihály Farkas from the Political and Central Committees of the Party, although this was not followed through consistently: on the following day Farkas was admitted into the Hungarian People's Army and then sent off to the Soviet Academy.[5]

The temporary improvement in international relations was reflected in a March 18 decree of the Presidential Council, ending the state of war against Germany, declared by the provisional government on January 20, 1945. That trend continued domestically when most victims of the show trials were gradually released, although their rehabilitation was extremely slow. A move in the opposite direction came on December 3, 1955, when Imre Nagy was expelled from the Communist Party by a decision of the Central Committee. By a quirk of fate, this almost coincided with the boost to Hungary's international reputation caused by the country's admission to the United Nations.

This very contradictory foreign and domestic situation in 1955, with alternating tension and détente, made it more difficult for the Hungarian People's Army to develop smoothly and evenly according to plan. The first part of the year was fairly quiet at the Miklós Zrínyi Academy, with training proceeding as usual, although its quality was jeopardized by shortcomings in the general and professional education of the officers. Statistics show that on February 1, 1955, 6,777 of them had completed six years of elementary school and 17,766 eight years, while 2,906 were high school graduates and 887 university graduates. Looking at their military training, 138 had completed the Soviet Academy, 157 the National Defense Academy and 16,895 a branch officers' school, while 887 had no military schooling at all. The saddest fact was that 95 percent of those in higher posts of command had only a civil school education (lower than high school graduation) and just 44 of them had completed a two-year or three-year academy course.[6] Although these figures show a substantial change since the beginning of the 1950s, it is also clear from them that the training was going to officers in lower ranks and posts, which promised a qualitative change in the near future. Another important aspect was the first Officers' Law, regulating the service of professional and reserve officers, which appeared on February 2.[7]

Yet the zigzags of domestic policy in the first half of 1955, the excessive load placed on officers, and the incessant transfers and housing trials brought looser discipline and a sharp increase in requests for a discharge. This prompted Colonel General István Bata, the minister of defense, and Lieutenant General Sándor Nógrádi, the chief of the Political Chief Section, to address an open letter to the officer corps on July 20, warning that "danger threatens internal order and discipline."[8] The causes that they cited included the appearance of anti-Party and right-wing ideas in the army, and the spur given to pacifist ideas by the easier international situation.

Feelings of security among officers (and so discipline) could also have been harmed by a government decision of September 1 to cut the strength of the People's Army by another 20,000.[9] They were aware that the problem of finding a job and reintegrating into civilian life was unresolved and that financial assistance to do so would be inadequate. Nor were officers reassured by the introduction of a small discharge payment and assistance with buying civilian clothes. During the same period, the Defense Command made some efforts to improve pay for professional officers. The Financial Group Command submitted a document to the minister of defense on "The Tendencies in Pay in the People's Army, the Party, and Other Areas of the People's Economy,"[10] recalling that there had been two pay rises in the Hungarian People's Army since 1950 (on December 1, 1951, and November 1, 1952), giving an average increase of 33.5 percent. This compared with an average rise of 67.2 percent over the same period for skilled and unskilled workers in industry. While a major general in command of a division earned 3,945 forints a month, a manager in industry received 3,400, a deputy manager 3,100 and a department head 2,800, with the chance of a monthly plan-fulfillment bonus of 1,000–1,500 forints a month. In fact the average monthly pay of the managers of three large engineering factories (Ganz, MÁVAG and Láng) was 5,950, 6,063 and 5,480 forints respectively. Officials in Communist Party and local government employment received far less: 3,350–3,800 forints for a Budapest Party secretary, 3,200–3,700 forints for a county Party secretary, 3,300–3,700 forints for a deputy chairman and executive secretary of the Capital City Council, and 3,100–3,600 forints for a county council chairman. These figures, largely unknown to most personnel, were more than offset by a further strength cut of 20,000 men, decided by the government on

October 18. In the event, 3,960 officers were discharged between September 1 and December 31, 1955, leaving a Hungarian People's Army strength on the latter date of 141,380 military and 8,751 civilian employees.[11]

The foreign and domestic political events and the consequent situation in the Hungarian People's Army formed the framework for military higher education in 1955. The 1954–1955 academic year still began in the National Defense Academy, but Colonel General István Bata, the minister of defense, formally bestowed on the institution the name of the Miklós Zrínyi Military Academy, in an order dated March 14, 1955.[12] This honor the complement could only celebrate in a brief break, before work continued. On March 23, for instance, Major General Endre Matékovits,[13] the commander of the Academy, reported to Major General Lajos Tóth,[14] the chief of General Staff, on experiences with the mid-year conference of the Correspondence Course in the 1954–1955 academic year.[15] Only three out of 52 students of the first-year course had been off sick during the March 8–18 event. According to the report, three students had been "not satisfactory" and five had not tried to do their duty as they had been unable to prepare. Another five who were "satisfactory" were so far below the requirements that this was not entered in their examination record. Only two persons were "not satisfactory" in the combat control tasks. Eight attempted it but fell too far below the desired standard for it to be entered in their examination record. The examinations could be resat up to May 1. Those who were far behind were referred to the Correspondence Faculty Command for a resit timetable. The combat control task was successfully completed by all.

The reports show in general that the vast majority of students had developed a lot, especially as "the Correspondence Course years include more than one student who began studying by correspondence with practically no previous combat training…" It was still registered as a problem that despite the order from the chief of General Staff, commanders were failing to check or even support the correspondence studies of their subordinates. Even now, the conference had only been successful because the commander of the Academy had expressly rejected all requests for leave of absence on service commitment grounds. But the many cases of inadequate preparation and absenteeism showed that the problem remained. The main trouble was that many commanders were sabotaging the minister of defense's regulation that allowed students a weekly day of study. "Commanders at various levels see the Correspondence Course as a necessary evil," ran one complaint from the commander of the Academy. He cited the Air Force Command as a bad example, as 90 percent of students from there were not satisfactory. This contrasted with the attitude in the Soviet Union, where "according to the Comrade Adviser… no commander would ever obstruct subordinates in their correspondence studies." This unworthy conduct by unit commanders caused six men to leave the first-year course.

According to the report on the preparedness of the Miklós Zrínyi Military Academy on April 1, 1955,[16] the institution had 285 generals and officers, 56 re-enlisted and 18 enlisted subalterns, 95 privates, and 476 officer students, 930 soldiers altogether, with a further 162 civilian employees. The branch and course structure of these was as follows:

Faculty	All-branch	Artillery	Armored	Air Force	Engineers	Signals Troops	Total
1st Year	67	35	20	18	13	12	165
2nd Year	56	37	19	16	15	14	157
3rd Year	28	35	18	19	13	10	123
4th Year	-	-	-	-	-	-	-
1-Year Higher Commanders' Course	20	11	-	-	-	-	31
Ministry of the Interior students (ÁVH)	25	-	-	-	3	3	31
Altogether	196	118	57	53	44	39	507

In charge of the state examinations at the Academy in 1955 was Major General István Kovács, head of the Operations Section of the General Staff.[17] He endorsed the Academy recommendations for this with the following differences:

- The number of Russian-language lessons should increase, in order to overcome shortcomings found in the teaching. The teaching aim was to impart mastery of military terminology, with the requirement that students should become capable in three years of making themselves understood "in a simple form."

- The marks for the combat oral examinations had to include a question each on the subjects of all-branch warfare, branch warfare, combat-technical data and staff work. Aspects of chemical warfare and nuclear defense also had to be added.

- The written examination for the armored students was to concern the activity of the armored division in the second, not the first, stage of the corps.

- The material for the anti-aircraft artillery students should include organizing the air defense of a Hungarian district, as well as the combat written examination.

- The Board chairman should be the chief of General Staff for the all-branch students and the branch commander for the students of the branches.

- One higher officer from the General Staff (or Branch Command) and one teacher delegated from the Academy had to be delegated to the examination boards.

- The oral and written material for the state examinations, taking into account the desires of the branch commanders, had to be submitted for approval by July 26, 1955.

In an order of April 26, 1955,[18] the minister of defense introduced an Academy badge for those who had obtained a diploma from the Miklós Zrínyi Academy or the Stalin Academy. In a further measure on April 5,[19] he allowed twelve Academy instructors to study at civilian universities (Miskolc, and the Loránd Eötvös University in Budapest); several of these became respected lecturers and department heads in subsequent decades.

Meanwhile, on April 18, the minister of defense had called for applications for the three-year Academy Course starting in 1955–1956,[20] stipulating the following, irrespective of rank:

- Candidates should have had appropriate previous training and at least two years' field experience;

- Candidates should have commanded or been the second-in-command of a company, battalion or brigade, or chief or deputy chief of staff of a battalion or division, divisional chief or second-in-command in a regiment or higher unit, in a field reconnaissance or officer-training or similar post, or brigade commander or deputy commander at a school as a teacher or department head, or on the staff of a higher unit or a central organization;

- Political officers should have done relevant work as a political deputy or in another political position;

- Candidates should have the requisite recommendations from their superiors;

- If possible, candidates should have completed middle school, or at least the eight grades of elementary school, or possess equivalent general education (this requirement can be ignored in certain cases) and have good intellectual capabilities;

- Candidates should be physically healthy.

Draconian decisions were sometimes taken to maintain discipline in the Academy, for instance an April 29, 1955, sentence of the Military College of the Supreme Court[21] on a student first lieutenant, who received six months' imprisonment and demotion for being absent without leave for 29 hours and handing his ID and pay book to an unauthorized person. In further efforts to raise the standard of military higher education, the government laid down in the fall of 1956[22] the requirement that at least 50 percent of the officer intake into the Academies had to be high school graduates or equivalent.

Returning to the call for applications mentioned earlier, the admission procedures for the basic three-year academic course were held in groups of 25–30 between July 15 and August 3.[23] Each lasted two days, one for professional and general examinations and the other for medical and personnel screening. The examination subjects were as follows:

- For all: social science, basic branch knowledge, topography, geography, history, Hungarian language and arithmetic;

- For air defense and signals officers: chemistry and physics;

- For air force officers: physics.

The minister of defense issued an instruction on July 14, 1955, on "improvement of the control and training work of the Ministry of Defense Central Bodies, military academies, training institutions, and units and higher units."[24] This defined the following main tasks for the Miklós Zrínyi Military Academy:

- Ending the teaching and training activity that was recondite, uselessly theoretical, and divorced from practicalities;

- Making the teaching more practical at every level, so that the students could arrange and carry out the teaching material in practice, in the spirit of the regulations;

- Paying special heed to the use of nuclear and chemical weaponry and night training;

- Handling as a special task improvement in methodological and scientific research work and in the training of scientific cadres;

- Having branch commanders and service chiefs examine at least every quarter their relevant Academy faculty and help with high-standard officer training, reporting their experiences directly to the minister;

- Ensuring that the Academy regularly received the most up-to-date military equipment;

- Providing the minister of defense by August 15 with the proposed complement table for the Miklós Zrínyi Military Academy in the fall of 1955 and including the Scientific Research Sub-department, the Correspondence and Regular Lecturers Sub-department, the Anti-Aircraft and Nuclear and Chemical Defense Faculties, the Academy Preparatory Course, the organizing regulations for reconnaissance officer training and the Academy Course;

- Holding command field exercises for the teaching staff from the first half of 1956.

The examination boards made the following recommendations in late July and early August for those beginning their Academy studies in the fall of 1955:[25]

Specialization	Recommended for Admission	Not Recommended for Admission
Infantry	40	14
Armor	14	7
Air Force	14	14
Engineers	13	3
Signal Troops	15	2
Logistics	9	2
Total	105	42

Other documents show that artillery and anti-aircraft classes also began in the 1955–1956 academic year,[26] with 18 and 14 admissions and 6 and 12 rejections respectively.

The instructions for conducting the end-of-year examinations soon followed.[27] On preparation days, reveille was at 05:45, with private study periods of 07:00–14:00 and 15:30–19:50, lunch and rest at 14:00–15:30, supper at 20:10–20:40, private study at 20:40–22:30, private free time at 22:30–23:00 and lights out at 23:00. The examinations were held at 07:00–15:00, and "survivors" could leave the Academy site at 16:00 with leave from the faculty commander. Professional cabinets to help with private preparation were open at 07:00–22:00 and faculty consultation services were available as well.

On August 27, Colonel General István Bata, the minister of defense, approved the proposed strength of the Miklós Zrínyi Military Academy, to take effect in the fall.[28] There would be 311 officers, 40 re-enlisted men and 3 enlisted subalterns, as well as 46 privates, making a total of 400 military posts. In terms of rank, it was allowed to have 1 lieutenant general, 3 major generals, 35 colonels, 136 lieutenant colonels, 114 majors, 20 captains, 1 first lieutenant and 2 lieutenants. The rank of lieutenant general for its commander (equivalent to a deputy chief of General Staff or section chief) was a special mark of respect for the institution, as were the ranks of major general for the three deputy commanders and colonel for 35 officers, mainly department and faculty heads. The complement included 482 regular, 40 course, and 38 ÁVH students.

The Organization and Mobilization Section issued a document signed by the minister of defense and the chief of General Staff on reorganizing the academies and training institutions,[29] stating the following: "The Miklós Zrínyi Military Academy is to continue operation with the approved strength, except that the hitherto separate Central Officers' Course at the Ministry of Defense is to be incorporated into it... From October 31, 1955, the special and mobilization courses will be subordinate to the commander of the Military Academy." This process of cessation and merger is remembered by Retired Colonel Dr. Lajos Móricz, once studies deputy to the commander of the Interpreters' Officers' School: "In the summer of 1955 there appeared a measure for the complete disbandment of the school, which suggested that there would not be any need for interpreter training in the future. 'A better place for preparing officers to be sent to the Soviet Academy will be the Academy, and the mobilization course, also brought under the school two years ago, must also be transferred to the Academy, where other [such] courses are operating,' ran the explanation from the authorities."[30] Although the organizational measure mentioned earlier would come into operation only on October 15, the new Nuclear and Chemical Training Faculty started work immediately.

At the end of September came the end-of-year examinations of the T (extension training) course. The following average marks were obtained.[31]

Class	Combat	Party History	Party Political Work	Artillery	Anti-aircraft	Military History	Class Average
T-1 (infantry)	4.40	4.40	4.50	-	-	4.30	4.40
T-2 (infantry)	4.60	4.24	4.38	-	-	4.23	4.36
T-3 (artillery)	4.33	4.11	4.44	4.22	-	4.44	4.33
T-4 (anti-aircraft artillery)	4.66	3.66	4.00	-	4.00	4.00	4.06
Average	4.49	4.11	4.34	4.22	4.00	4.24	

The report on the status of the Academy dated October 1, 1955,[32] contained the following statistic: the regular strength comprised 289 generals and officers, 68 re-enlisted and 9 enlisted subalterns, 99 privates, and 520 students.[33] This gave a military complement of 965 (up by 43) and a civilian workforce of 162 (down by 133). However, the Academy was under strength by 6 officers, 15 re-enlisted subalterns, 1 private, and 48 students on October 1 and over strength by 9 enlisted subalterns and 5 ÁVH students.

According to the Academy Command's statement of October 13, 1955,[34] there were 129 state examinees, of whom 122 received Academy diplomas. As mentioned before, the end-of-year examinations of the first-year Academy Course were held on October 8–11. The average score of 4.09 can be considered quite respectable.

On October 13, 1955, the commander of the Academy issued an order[35] on the passing-out ceremony, in which he emphasized the following fact: "Our Academy, for the first time in the history of our People's Army, is producing diploma graduates who have completed a three-year course and will receive the Academy badge." The diplomas and the badges for the Miklós Zrínyi and Stalin Academies were to be received at a joint ceremony in the presence of the chief of General Staff. The minister of defense would give a supper in honor of the graduates in the ceremonial hall of the Miklós Zrínyi Academy. He was careful to add the following: "On the occasion of the passing out of our third-year students with a diploma and a badge, our Academy will also receive the honor of having the leaders of our Party and Government present." So security measures came to the fore. From the guard change to the guard change on October 15, there was to be a kitchen duty guard, and on October 14, the task had to be performed by the head of the Military Logistical Faculty himself. On October 13–14, the cold store had to be closed and sealed by a board consisting of the duty officer, the kitchen duty officer, the head of the medical service, the head of military supplies and the head of stores, and the duty officer had to check several times that the locks and seals were untampered with. The medical officer had to be present personally on October 14 while the dishes were being prepared, and the duty officer was to be reinforced by a deputy.

Colonel General István Bata, the minister of defense, greeted the graduate students in an order dated October 14,[36] underlining the following:

From the Academies of the Hungarian People's Army issue for the first time today officers who have completed a three-year academy [course]… Let us never lose sight of the fact that the difficult and responsible tasks awaiting the officers of our People's Army can only be carried out in full by those who keep developing the knowledge that they have gained… Let their work be marked by a high level of meticulousness and a feeling of responsibility for their subordinates and the training of Hungarian youth. Let them be exemplary in carrying out their tasks to the full.

The November 3 admission examination for the first year of the Correspondence Course was covered in an order of October 25 from the commander of the Academy.[37] The examination for the Academy teachers differed from the "normal" examination only in that they were exempted by the commander of the Academy from the examinations in their own subjects and in topography. The second-year Academy students also took end-of-year examinations during October.[38]

The commander of the Correspondence Faculty submitted a report on its activity[39] as assistance for the commander of the Academy's assessment of the 1954–1955 training year. It states that all the correspondence students, apart from a few laggards, had covered the complete syllabus. The half-year examination averages were 4.12 for the first years and 4.15 for the second years. The level of student work had improved, as had their use of command language. They had also shown good results in the social science subjects. However, there was a problem in that certain commanders were not making available to students the free time prescribed in the minister's order, or not supervising their subordinates' study activity. Based on these circumstances, it was requested that the commander of the Academy seek to ensure that branch commanders and the chief of staff take the necessary measures in this respect.

Based on experiences in the 1954–1955 academic year and with the state examinations, it could be concluded that the Tactics/Strategy Faculty had successfully resolved intricate new problems such as translating the principles of the new Combat Regulations into suitable teaching material, revising the thematic programs, and adding new detail to them, and beginning to devise principles for the use of nuclear weapons and to incorporate them in teaching. However, the members of the state examination sub-committee considered (while emphasizing the good results) that "much greater attention should be paid in future to logical explanation and justification of the general combat principles and methods of implementation. We recommend, as a way of doing this, increasing the amount of practical work, making field activities more frequent, making more thorough technical preparation for staff command exercises, and carrying these out in a more practical way."[40] The Artillery Faculty also underwent major qualitative development during the year: two of the eight new teachers had completed the Soviet Military Academy and two the Hungarian one, while one had experience of commanding an artillery brigade. This meant that 19 of the 27 teaching officers in the fall of 1955 had academy qualifications.[41] Another important event for the future was the integration of the Interpreter Officers' School (by then called the Ministry of Defense Central Officers' Course) and the mobilization and recruitment courses operating alongside it into the Miklós Zrínyi Academy as the Special Faculty.[42] An important development at the Air Force Faculty was the arrival in the spring of 1955 of Captain of the Air Force József Szabó, who had just graduated from the Soviet Air Force Academy, as well as another teacher and another specialist instructor. (Szabó would later be a lieutenant general, head of the air force of the Hungarian People's Army, a doctor of military studies, a university professor and president of the Hungarian Military Studies Society.)[43]

Those examinations and reports essentially brought to a conclusion the 1954–1955 academic year at what had started as the National Defense Academy and had become the Miklós Zrínyi Military Academy. The institution can be said to have performed its task successfully under a complicated set of conditions.

Notes

1. On Rákosi, Nagy and Farkas, see the Biographies of Key Personalities.
2. Jenő Gergely and Lajos Izsák, *A huszadik század története* [History of the Twentieth Century] (Budapest, 2000), p. 353.
3. *Ibid.*, p. 354.
4. On Hegedűs and Gerő, see the Biographies of Key Personalities.
5. László Csendes and Tibor Gellért, *Kronológia a Honvédség történetéből 1945–1990* [Chronology of the History of the *Honvédség*, 1945–1990] (Budapest, 1993), p. 148; Hadtörténeti Levéltár, Magyar Néphadsereg [Military History Archive, Hungarian People's Army] (hereafter cited as HL MN) 1955/T–Titkárság [Secretariat]. 3/2. őrzési egység [guard unit] (hereafter cited as ő. e.) (April 16, 1955). Ministry of Defense Officer's Order No. 9.
6. Csendes and Gellért, *Kronológia a Honvédség történetéből 1945–1990*, p. 145.
7. *Ibid.*, p. 147.
8. *Ibid.*, p. 151.
9. *Magyarország hadtörténete* [Military History of Hungary], editor-in-chief Ervin Liptai (Budapest, 1985), p. 554.
10. HL MN 1956/T. 2. ő. e. (March 9, 1955), ff. 43–47.
11. Csendes and Gellért, *Kronológia a Honvédség történetéből 1945–1990*, pp. 151, 144 and 152.
12. HL MN 1955/T–Titkárság 3/2. ő. e. (March 14, 1955).
13. See the Biographies of Key Personalities.
14. There were two senior officers of that name: Major General Lajos Tóth, the chief of General Staff, and Colonel, later Major General Lajos Tóth, the head of the Organization and Mobilization Section.
15. HL MN 1955/T–94/2. ő. e. (March 23, 1955).
16. *Ibid.*, T–94/1. ő. e. (April 1, 1955).
17. *Ibid.*, T–94. ő. e. (April 20, 1955).
18. Csendes and Gellért, *Kronológia a Honvédség történetéből 1945–1990*, p. 148.
19. HL MN 102/03–405. ő. e., ZMKA [Miklós Zrínyi Military Academy] Order No. 99 (April 26, 1955).
20. HL MN 1961/T–33. ő. e., ZMKA Order No. PK 037 (May 10, 1955).
21. *Ibid.*, T–33. ő. e. (May 10, 1955).

22. *A magyar állam szervei 1950–1970. Központi szervek* [Bodies of the Hungarian State, 1950–1970. Central Bodies] (Budapest, 1993), p. 372.
23. HL MN 1961/T–33. ő. e., ZMKA Order No. PK 056 (July 1, 1955).
24. *Ibid.*, T–Titkárság, 3/2. ő. e. (July 14, 1955).
25. *Ibid.*, T–94/1. ő. e. Minutes of July 16, 21, 26 and 28, and August 2, 1955 (ff. 37, 40, 42, 42/a, 46, 54 and 57).
26. HL MN 1955/T–94/1. ő. e. (Aug. 1, 1955), ff. 50 and 52.
27. HL MN 1961/T–33. ő. e. (Aug. 10, 1955).
28. HL MN 1967/T–167. ő. e. (Aug. 27, 1955).
29. *Ibid.*, T–163/I. ő. e. (Sept. 22, 1955).
30. Dr. Lajos Móricz, *Békében és háborúban. Egy magyar katonatiszt emlékei a XX. század nyolc évtizedéről* [At Peace and at War: A Hungarian Officer's Memories of Eight Decades of the Twentieth Century] (Budapest, 2000), p. 174.
31. HL MN 1955/T–94. ő. e. (Sept. 30, 1955).
32. *Ibid.*, T–94/1. ő. e. (Oct. 1, 1955).
33. The ÁVH students were not included.
34. *Ibid.*, T–94. ő. e. (Oct. 13, 1955), f. 38.
35. HL MN 1961/T–33. ő. e. (Oct. 13, 1955).
36. HL MN 1955/T–Titkárság, 3/2. ő. e. (Oct. 14, 1955).
37. HL MN 1961/T–33. ő. e. (Oct. 25, 1955).
38. HL MN 1955/T–94. ő. e. (Nov. 1, 1955).
39. *Ibid.*, T–94. ő. e. (Nov. 26, 1955).
40. Lieutenant Colonel József Varga and Lieutenant Colonel (ret.) Lajos Kovács, "Összfegyvernemi harcászati tanszék" [The All-Branch Combat Faculty], *Akadémiai Közlemények. Az Akadémia szerveinek története* III/A. *A Szárazföldi ágazat története.* 120 (1985) 2: 40–41.
41. Colonel Dr. Gyula Enzsöl, "Rakéta és tüzér tanszék" [The Missile and Artillery Faculty], *Akadémiai Közlemények. Az Akadémia szerveinek története* III/A. *A Szárazföldi ágazat története.* 120 (1985) 2: 106.
42. Lieutenant Colonel Dr. Boldizsár Bakity and Lieutenant Colonel Dr. Antal Mátyási, "Szervezési, hadkiegészítési és igazgatási tanszék" [The Organization, Mobilization and Administration Faculty], *Akadémiai Közlemények. Az akadémia szerveinek története* III/B. *Szárazföldi ágazat tanszékeinek története.* 120 (1985) 3: 145.
43. Lieutenant Colonel Dr. Jenő Bartos, Lieutenant Colonel (ret.) Lajos Kovács and Lieutenant Colonel (ret.) József Hajdó, "Repülő hadműveleti-harcászati tanszék" [The Air Force Tactics and Strategy Faculty]. *Akadémiai Közlemények. Az akadémia szerveinek története* IV. *A Légvédelmi és repülőágazat története.* 120 (1986) 4: 25.

Chapter XI

The Miklós Zrínyi Military Academy in 1956:
Political Events and Military Conditions

International political events in 1956 continued to display conflicting tendencies, as they had in 1955 and the previous Cold War years. Signs of détente and mutual concessions between the Great Powers alternated with heightening tensions and retaliation. Those important events that promoted détente included the following:

- The Twentieth Congress of the Communist Party of the Soviet Union (February 14–25);
- Morocco and Tunisia gained independence from France (March 2 and 20);
- Austria joined the Council of Europe (April 16);
- The Soviet Union announced that it was cutting its armed forces by 1.2 million (May 15);
- The evacuation of the Suez Canal Zone by Britain and France was completed (June 13);
- The Central Committee of the Communist Party of the Soviet Union adopted a resolution "On Overcoming the of Personality and Its Consequences" (June 30);
- Japan and the Soviet Union issued a statement ending their state of war and establishing diplomatic relations (October 19);
- The Soviet Union protested over the Anglo-French-Israeli aggression against Egypt (October 31);[1]
- The last occupation forces left Port Said (December 22).

Many political and military measures taken *impeded détente*, for instance the following:
- The establishment of East Germany's People's Army (January 18);

- An uprising against the Communist regime in Poznań, Poland (June 28);

- President Nasser of Egypt announced the nationalization of the Suez Canal Company (July 26);

- The Revolution and War of Independence in Hungary (October 23–November 11);

- Israel attacked Egypt (October 29);

- A statement on developing friendship and cooperation between the Soviet Union and the other socialist countries and reinforcing the basic principles of these (October 30);

- An Anglo-French ultimatum calling on Egypt to withdraw its forces from the Suez Canal Zone (October 30), followed by several days' bombardment and the landing of British and French troops at Port Said (November 6);

- The yacht *Granma*, carrying revolutionaries led by Fidel Castro, landed in Cuba's Oriente province to start guerrilla warfare against the Cuban regime (December 2).

These cardinal international events had similarly mixed effects on Hungary's domestic political relations as well. One of the greatest from several points of view was the aftermath of the Twentieth Congress (February 14–25) of the Soviet Communist Party. The proceedings at the closed session, including Khrushchev's four-hour denunciation of Stalin's crimes and the cult of personality, soon leaked out, and most of the public were expecting the Hungarian Communist leadership to draw far-reaching conclusions from them. These expectations gave way to consternation when Mátyás Rákosi[2] delivered a speech on March 12–13 to his Party Central Committee claiming that the Soviet Twentieth Congress had largely vindicated his Party's policies. Although this assessment was criticized by János Kádár, Márton Horváth, Zoltán Szántó and Erik Molnár,[3] the report was passed unanimously by the Central Committee. However, this only activated the Reform-Communist intelligentsia further. The first major event organized by the reformist Petőfi Circle took place on March 17. The mounting anti-Stalinism prompted the government to warn the Soviet leadership of the dangers of the policies that Rákosi espoused. Prime Minister András Hegedűs,[4] for instance, informed Soviet Ambassador Yuri Andropov on May 4 that "Rákosi has dealt little recently with the important political and economic problems, and sometimes omits to pay sufficient attention to the urgent tasks facing the Party."[5] Andropov heard similar accounts from other Communist leaders. These interventions, coupled with pressure from the Political Committee, may have urged Rákosi eventually to admit responsibility for fostering the cult of personality and for illegalities committed by the regime, speaking at a meeting of Budapest activists on May 18. His belated self-criticism appeared for a time to assure Rákosi of the full confidence of the Soviet Presidium, as Mikhail Suslov, the "chief ideologist" of the Soviet Communist Party, assured him during a visit on June 7–14.

But the gain was a short one. Yugoslav President Josip Tito announced during a summit in Moscow at the end of June that he was not willing to reach a rapprochement with Rákosi, and at the beginning of July Ernő Gerő[6] informed Andropov (who may have felt it already) that the situation in Hungary was "very serious"[7] and that it might be better if the Soviet leadership did not insist so strongly on Rákosi. As a result, it was decided at the Central Committee meeting on July 18–21, with the active cooperation of Anastas Mikoyan, who had been sent to Budapest to gain "local information," that Rákosi would be relieved of his position as Party first secretary and his membership of the Political Committee and would be succeeded by Gerő. On the following day, Rákosi and his wife left Hungary for good. The same meeting elected János Kádár as a Central Committee secretary and expelled Mihály Farkas from the Party,[8] recommending to the Presidential Council that he be stripped of his military rank.

Hungarian society became irresistibly radicalized in the late summer and early autumn months. As the situation became increasingly tense, László Rajk,[9] Lieutenant General György Pálffy, Tibor Szőnyi and András Szalai were reinterred on October 6, followed on October 13 by four army generals executed after show trials: László Sólyom, Gusztáv Illy, György Pórffy, Kálmán Révay and István Beleznay. These events may have contributed to the fact that in the final days before the Revolution, Imre Nagy was also rehabilitated and readmitted into the Communist Party.[10]

This very contradictory foreign and domestic political situation, with its alternation of tension and détente, may well have impeded or even prevented any even, planned, balanced development of the Hungarian People's Army in 1956. On January 1, 1956, there were 26,231 serving officers in the army, 30 of them generals. Efforts had been underway since September 1955 to improve their pay further. For instance, Colonel General István Bata,[11] the minister of defense, wrote a letter on February 2, 1956, to Mátyás Rákosi, as president of the National Defense Council, on the subject of "the pay system of certain ranks in the People's Army,"[12] reporting that he had carried out the review of pay, supplements and bonuses based on his August 1955 report and established that the pay in the new posts had not been regulated during the many reorganization efforts, that the 80 percent supplements permitted in 1949–1951 were no longer justified, and that abolishing them could save 8 million forints, of which 6.5 million might be diverted for a pay rise for the commanders of "the sub-units and units (squads and companies) that form the backbone of the army." The minister of defense was trying to circumvent the National Defense Council by arranging matters directly with Rákosi. On February 10, he sent a largely similar letter to Prime Minister András Hegedűs, emphasizing the point that his proposal would make no difference financially: "In fact, 1.5–2.0 million forints have been saved; it would be desirable for me to issue the order for payment of the new salaries on March 1." On the following day Hegedűs wrote in the letter "I agree." The minister of defense's efforts were successful. On April 1, 1956 (a month later than intended), an order was issued on "Implementation of the 'Enacting Order' on the wage scale of the personnel of the Hungarian People's Army and settlement of other financial allowances."[13]

What decisively influenced the thinking of the majority of society and the army was this statement in the Twentieth Congress resolution: "War is no longer fatally inevitable: there is a realistic chance of preventing the third world war."[14] This doctrine had numerous positive and negative effects on the Hungarian People's Army. It may also, along with other extremely important statements at the Twentieth Congress, have played a part in the decision to place the hitherto exclusively Party-controlled Armed Forces and Organizations under state control, through a government decision of April 19, in the fact that Hungarian People's Army engineers began to remove the technical barriers along the Hungarian–Austrian border on May 10, and the government's decision on August 1, based on the July resolution of the Communist Party Central Committee, to reduce the strength of the Hungarian People's Army by 15,000. In addition to the last, the government also announced at the July session of Parliament that it intended to spend 35 percent less on defense in the following year than it had in 1953.[15]

Decisively important to basic and higher-level officer training was an order from the minister of defense on June 19, 1956, on reorganization of the military academies and officers' schools.[16] The measures affecting the *military academies* were the following:

- The training period at the Stalin Military Political Academy and in the Signals and Engineering Faculties of the Miklós Zrínyi Military Academy would be four years from December 1, 1956.

- The training period in other Miklós Zrínyi Military Academy faculties would remain at three years. The officers presently in the third year at either of the Academies would graduate under the present system in 1956.

- The training would begin each year from 1956 on December 1.

- Officers graduating from three-year faculties would pass out by April 10 each year from 1958, and those graduating from four-year faculties by September 30. From the fall of 1956, the Correspondence Faculty of the Miklós Zrínyi Academy would offer only all-branch training.

- The Petőfi Military Political Officers' School would continue as a two-year extension course of the Stalin Military Political Academy.

- The measures affecting *officers' schools* were drastic and confusing, but yielded savings:

- The Dózsa Infantry and Rákosi Armored and Motorized Officers' Schools would be combined under a single command.

- The Kossuth Artillery and Áron Gábor Artillery Engineering Officers' Schools would merge at the latter's base and in part of the János Kiss Barracks, and continue to train officers in artillery, anti-aircraft artillery, and artillery and radio engineering.

- The Zalka Signals and Táncsics Engineering Officers' Schools would merge at Szentendre and continue to train officers in engineering, signals, chemical defense and warships.

- The status of the Kilián Air Navigation and Vasvár Air Force Officers' Schools would be unchanged.

- The Savaria Military Supply Officers' School would close and training of the third year would continue until its completion in the Military Supplies Faculty of the Miklós Zrínyi Academy.

- The officers' examinations would be completed by August 31, 1956, and the officers' inauguration would be held by September 5.

However, uncertainty and dissatisfaction in society and in the armed forces began to deepen again as leadership divisions prevented action on the reassuring elements in the July 1956 resolution of the Communist Party Central Committee. This applied particularly to the fact that most officers changed posts and stations almost every year, so that some 20 percent of them were obliged to live separately from their families. The sense of a steady decline in the political morale of the Hungarian People's Army prompted assemblies at larger garrisons, addressed by the president of the Presidential Council, members of the government and leading army commanders, where they managed to answer many questions, but failed to answer many others. When the Chief Prosecutor's Office took Mihály Farkas, the former minister of defense, into custody on October 12—one day before the reburial of the generals convicted in show trials—it was a move designed to relieve some of the mounting tension.[17]

There may also have been a connection with the mounting national feeling when Colonel General Bata, the minister of defense, addressed a letter on September 22 to his Soviet counterpart, Marshal Zhukov, recommending that 34 adviser posts be abolished and 6 new posts created, so reducing the number of Soviet military advisers in the Hungarian People's Army to 54.[18]

The defense leadership called a meeting in Budapest for October 22–24, 1956, for officers of the Hungarian People's Army from regimental commanders and their political deputies upward. There would have been plenty to discuss: the new training year was just beginning; units were being disbanded due to the troop cuts; 12,322 officers were demobilized between November 15, 1953, and October 23, 1956, including 2,633 between January 1 and October 23, 1956, alone; the stringent war service regulations had been withdrawn; morale and discipline in the army had deteriorated drastically.[19] However, the rapidly rising tensions in the country prompted the authorities to cancel the meeting, although several commanders were unable to return to their troops during the critical period.

On the evening of October 23, the Communist Party Central Committee decided that "an armed attack on the people's power had begun" and that this had to be met with armed force. However, the military forces stationed in Budapest consisted of only the Miklós Zrínyi Military Academy, the Sándor Petőfi Military Political Academy and some branch officers' schools, and so the defense leaders advised the Political Committee to bring in provincial troops and to ask the Soviet forces for a show of strength as well. On October 24, the Central Committee created a Military Committee with absolute powers to direct the suppression of the "counter-revolutionary rebellion," appointing to it Antal Apró,[20] Lajos Fehér,[21] László Földes,[22] Major General István Kovács and Imre Mező.[23] On their recommendation, martial law was declared, with an amnesty for those laying down their arms by a deadline.

The military leadership failed to issue clear orders to the troops. The commonest order was to "act at your own discretion" and they were forbidden to use weapons even if attacked. Often they would have had no means of doing so in any case; tanks brought up from the provinces were not allowed to stockpile ammunition, for instance. The hesitancy is unsurprising, seeing that vacillating decisions from the country's leadership were being carried out by an incompetent ministry. According to a report on educational attainment by the Personnel Chief Section in the autumn of 1956, "The situation is worst in the M[inistry] o[f] D[efense]… 38.7 percent of MOD officers have completed a higher military school… 10.3 percent have completed six elementary grades, 44.9 percent eight elementary grades or four secondary grades; 24.9 percent have graduated from high school, and 19.9 percent from university or college."[24]

At 17:00 on October 23, 1956, Colonel Endre Pesti, the commander of the Academy, ordered a meeting of the personnel present at the Sándor Petőfi Political Academy. In the evening hours, the commander of the Academy received a telephone call from Colonel

General István Bata, the minister of defense, inquiring, in the light of the situation in the city, whether he could rely on the Academy. The answer was affirmative, and not long afterwards an order arrived from the Ministry of Defense calling for 150 men from the Petőfi Academy, each with a bayonet and rifle and 40 rounds, to strengthen the defense of the Radio building, to restore order, and to ensure that the Gerő speech was broadcast. The minister of defense gave Colonel Pesti express orders: "It is forbidden to fire on the people!" But this caused indecision and chaos among the volunteers for the task. Since the Academy did not possess vehicles for transporting troops, there was over an hour's delay before they could depart for the Radio building, a few minutes after 20:00. Then they ran into a large, angry crowd on the corner of Múzeum körút and Bródy Sándor utca. The commander of the detachment, Lieutenant Colonel István Trizna, training deputy to the commander of the Academy, was assaulted, and the crowd demanded that they turn back. This they were obliged to do, because of the order forbidding the use of weapons, to the sounds of applause and cheering from the crowd.

This incident was reported to the Ministry of Defense and the chief of General Staff, prompting a further order to carry out the task, while retaining the prohibition on firing on the people. The detachment tried to overcome the inconsistency of this by concealing pistols and tear-gas and hand grenades in their pockets and set off for the Radio building again behind the national tricolor, singing songs of the 1848 Revolution. The Ministry of Defense staff promised to inform the defenders of the building of this, to ensure that they were admitted, but this was not done. Instead, the Academy students were met by warning fire, tear-gas grenades and hoses, and only 35–40 of them gained entry. The rest returned to the Academy or mingled with the crowd.[25]

On October 25, the students of the Kossuth Artillery Officers' School were involved in heavy fighting, with serious losses on both sides. On that day, Colonel Pál Maléter[26] arrived with five tanks at the Kilián Barracks on the corner of the Nagy körút and Üllői út, and was involved in an armed incident with the Corvin köz insurgent group. Lieutenant General Károly Janza took over as head of the Ministry of Defense.

The two bloodiest days in the Revolution were October 25 and 30, which are worth comparing. On October 25, a huge unarmed crowd gathered in Kossuth tér, demanding the removal of Ernő Gerő. The demonstrators began to fraternize with those guarding the Parliament, including the crews of some Soviet tanks that were just arriving to reinforce them. They jumped up on the tanks and stuck Hungarian flags down their barrels, to which the Soviet crews responded in a friendly fashion. This fraternization enraged General Serov, the commander of the Soviet GPU (secret police), and the members of ÁVH units hidden on the roofs of nearby buildings.[27] It still remains unclear whether the tanks commanded by Serov or the ÁVH men on the roofs opened fire first, but both blazed down on the peaceful, unarmed demonstrators, with murderous results. More than 100 demonstrators were killed.

The ÁVH men guarding the Communist Party headquarters in Köztársaság tér on October 30 took several freedom fighters prisoner but later released them again. But the news spread that some of them were still in ÁVH hands. The insurgents outside wanted to enter the building peaceably, in order to free their imprisoned comrades, or to satisfy themselves that there were no prisoners in the building, before departing peaceably again. But the peaceful entry was met by murderous fire from the ÁVH side. Violence was then met with violence, but the insurgents could not know they had been denied entry because a Bolshevist staff colonel and a few political officers were organizing a counter-revolutionary coup designed to overturn the government of Imre Nagy.

The crews of the tanks sent from the provinces to help the Bolshevists, unfamiliar with the area and confused about the situation, joined the insurgents instead of the counter-revolutionaries, so tipping the "battle" in their favor. The freedom fighters were incensed by the mindless violence of the Bolshevists and slaughtered about a dozen ÁVH men. By lynching Bolshevists who had surrendered, the insurgents made a grave mistake, but the losses on both sides of about 30 in Köztársaság tér were small compared with the bloodshed in Kossuth tér on October 25, for instance. The former were reacting to unprovoked fire, which had cost them as many lives as they took. The Bolshevists in Kossuth tér had used tank fire against peaceful, unarmed demonstrators. Yet the propagandists of the Kádár regime were to ensure that the Köztársaság tér incident would become known worldwide, while the Kossuth tér massacre was hushed up as far as possible.

Fighting also intensified elsewhere in the country on October 26. The county divisional premises of the ÁVH were attacked in Pécs, as was the anti-aircraft artillery in Sztálinváros (today's Dunaújváros), where several people were killed and wounded. Ten civilians lost their lives and four border guards were lynched outside their barracks in Mosonmagyaróvár. A crowd attempting to obtain weapons from the nearby Győr Border Guard's barracks was fired on, producing several dead and injured. Major General Lajos Gyurkó, army corps commander of Kecskemét,[28] sent aircraft against unarmed protesters in Kiskőrös, Kecskemét, Tiszakécske and Csongrád. Ten people in a crowd demanding the release of prisoners in Miskolc were mowed down by fire and the crowd lynched seven police and ÁVH. On the same day, Lieutenant General Károly Janza, the new minister of defense, ordered "the military units deployed against the armed groups to continue annihilating these and restoring order."[29]

In a joint order on October 28, Károly Janza and Dr. Ferenc Münnich, the interior minister, ordered the state arms on uniform caps to be replaced by a ribbon in national colors and the greeting to be *bajtárs* (comrade-in-arms) again, not the communist *elvtárs* (comrade).[30] At 13:20 on the same day, Prime Minister Imre Nagy ordered an immediate cease-fire. He announced the abolition of the State Security Office (ÁVH) and said that the Soviet troops would return to barracks at his request. The Revolution had ostensibly won. Two days later, the Military Committee chaired by Antal Apró ceased to function. On the same day, Minister of Defense Janza ordered the establishment of Revolutionary Military Councils to promote "revolutionary discipline." Their tasks included "discussing and advising on every important order and action by commanders appointed by the minister of defense... If the majority of members of the Military Council do not agree with the orders of the commander, the commander shall report to his superior commander."[31] On October 30, Imre Mező, first secretary of the Budapest Party Committee, along with Colonels József Papp and János Asztalos, died in the siege of the Budapest Party headquarters just described, and Colonel Lajos Tóth[32] was seriously wounded. The latter were those officers who had earlier been present in the Communist Party headquarters for the purpose of forming a counter-revolutionary terror organization. Also on October 30, the Revolutionary Armed Forces Committee of the Republic was established, under the command of Major General Béla Király, to provide overall direction of the army, the police and the National Guard. On that day, Colonel Pál Maléter was appointed as deputy to the minister of defense, and Béla Király was made military commander of Budapest; Király was made commander-in-chief of the National Guard two days later. At the same time, marks of rank worn in the armed forces were moved from the shoulder to the lapel and a start was made to supplying traditional Bocskai caps similar to a kepi, to replace Soviet-style military headgear. Maléter, by then a major general, became minister of defense in the coalition government formed on November 2.[33]

On the night of October 30–31, Soviet forces entered the country in numbers approaching those of the Allied forces in the 1944

Normandy landings. Their systematic moves to overthrow the democratic Hungarian government amounted to a war by the Soviet Union against Hungary. Since the Revolution had not swept away the basic institutions of socialism, this can be called the first war between two socialist countries.[34] A revolution is a domestic affair: progressive forces overthrow an obsolete political, social or economic system and replace it with one that matches their ideology. That is a victorious revolution, and the Hungarian Revolution of 1956 had won. The new democratic Hungarian society and government were trampled by outside power in a war, not by domestic forces. The armed Soviet intervention turned the Revolution into a war of independence, in a process that can be described briefly as follows.

On October 25, 1956, the Central Committee of the Hungarian Workers' Party was reconstituted, with János Kádár at its head, and the prime minister formed a new government. On October 26, Imre Nagy stated in a radio speech that his government had started negotiations on the withdrawal of Soviet forces. On October 28, direction of the Hungarian Workers' Party was taken over by a six-person presidium headed by Kádár, and the Central Committee termed the events that were occurring a "national democratic revolution."[35] On the same day, Prime Minister Imre Nagy had talks with Soviet Ambassador Yuri Andropov and called a cease-fire at 13:20. Also on October 28, the UN Security Council placed the situation in Hungary on its agenda. At dawn on October 31, as mentioned, vast Soviet forces entered the country and began military operations against it, without any declaration of war. On November 1, Imre Nagy summoned Andropov and demanded the immediate withdrawal of the Soviet forces. Then at 19:50 he made a radio speech, announcing, among other things, that "the Hungarian national government, moved by a deep sense of responsibility for the Hungarian people and its history and expressing the undivided will of millions of Hungarian people, declares Hungary's neutrality and its withdrawal from the Warsaw Pact."[36] The Soviet ambassador was informed of this in writing, and UN Secretary-General Dag Hammarskjöld in a telegram. János Kádár, in a recorded speech on Hungarian Radio, emphasized the point that there had been a "glorious uprising" on October 23 and the struggle for its democratic achievements was continuing.[37]

The prime minister reshuffled his government again on November 3. At midnight, General Ivan Serov, the commander of the KGB, arrested the Hungarian delegation engaged in negotiations with the Soviets at their base at Tököl. On the following day, the violent phase of the war against Hungary began. In the early hours of the morning, the prime minister made a historic announcement:

Imre Nagy, the president of the Council of Ministers of the Hungarian People's Republic, is addressing the Hungarian people:

This is Imre Nagy speaking, the president of the Council of Ministers of the Hungarian People's Republic. Today at daybreak Soviet troops attacked our capital with the obvious intention of overthrowing the legal Hungarian democratic government.

Our troops are in combat. The government is at its post. I notify the people of our country and the entire world of this fact.[38]

Almost at the same time, a recorded speech by János Kádár was being transmitted from Szolnok, stating that a Revolutionary Worker-Peasant Government had been formed and had begun to eliminate the *counter-revolution* with the help of the armed forces of the Soviet Union. Dr. Ferenc Münnich, the deputy prime minister and minister for the armed forces,[39] ordered the units of the Hungarian People's Army not to open fire on the Soviet troops, but to send negotiators to treat with them. Shortly after this dramatic

radio announcement, Imre Nagy and some of his leading colleagues—with some family members, 43 persons altogether—fled
to the Yugoslav Embassy, where they received asylum.

In a radio speech on November 11, János Kádár stated as head
of the Hungarian Socialist Workers' Party and the government that
the uprising had been suppressed. This terse announcement, wish-
ful thinking to some extent, was true insofar as the major armed
battles were over, as Major General Béla Király,[40] commander of
the National Guard and of the Budapest garrison, had retreated into
the Buda Hills on November 4, and after a clash on November 11,
near Nagykovácsi, retreated into exile before decisively superior
forces: "It became increasingly clear that the Kádár regime was
staying, whereas we still knew nothing about where Imre Nagy
was. Under the circumstances, there was no sense in still maintain-
ing the High Command of the National Guard. I dissolved the staff
and we carried on towards Austria as armed refugees,"[41] the general recalled. Nonetheless, there were sporadic shooting incidents in
later weeks and protests of various sizes, so that consolidation
seemed a very long way ahead. On November 14, the Greater
Budapest Central Workers' Council was formed; it demanded the
withdrawal of Soviet troops, restoration of the Imre Nagy government, and an institutional guarantee of political democracy, while
condemning all attempts at restoration and pledging faith in the
underlying principles of socialism, including state and social own-
ership. János Kádár, who negotiated with the Workers' Council del-
egates on that day, stated that he did not see Imre Nagy as a revolutionary and did not wish to restrict his freedom of movement: "It
depends only on him whether he takes part in political life."[42]
Unfortunately, that undertaking was not kept. Nor was the
Yugoslav–Hungarian intergovernmental agreement, signed on
November 21, under which the Hungarian government guaranteed
that the refugees in the Yugoslav Embassy could return to their
homes without prosecution. Instead it was agreed, probably on the

following day, with the Romanian First Secretary Gheorghiu-Dej and Prime Minister Shivu Stoica, who were conducting talks in Hungary, that Nagy and his associates would be deported to Romania. Their bus, which left the Yugoslav Embassy at 18:30 on the same day, was soon stopped by Soviet officers, who released the diplomats and sent the politicians and their families to the KGB Command in Mátyásföld. The following day, after Imre Nagy refused to go abroad voluntarily and Dr. Ferenc Münnich had no success offering the prisoners individually the chance to change sides, they were flown to Romania. This fact was announced in a distorted form in an official Hungarian communiqué on November 24: "Imre Nagy and his associates asked the Hungarian government over two weeks ago for permission to leave the territory of the Hungarian People's Republic for that of another socialist country. By agreement with the Romanian People's Republic, Imre Nagy and his associates left for the territory of the Romanian People's Republic on November 23."[43]

The comedy continued. Kádár announced the following in a radio speech on November 28: "We have given a promise not to start criminal proceedings for their past serious crimes, of which we too have learnt only latterly. This we will keep; we do not consider their departure to be for all time."[44] If the prime minister did think that this was meant seriously, such illusions would soon have been shattered, for Suslov, Malenkov, Aristov and Serov, then in Hungary, agreed with Kádár and Münnich on November 30 that the mechanism of reprisals would be switched on very soon. Indeed, between six and eight revolutionary leaders, including János Szabó of the Széna tér group, "were to be put through extraordinary proceedings and executed as examples."[45]

Budapest and the provinces underwent a wave of strikes and protests from early November to mid-December, with some counter-demonstrations, and crowds raked by gunfire in some places (e.g. Salgótarján, Miskolc and Eger) causing deaths in double figures. Thereafter, the situation gradually returned to normal. Some sources estimate that about 2,500 people lost their lives between October 23, 1956, and January 1957; 44 percent were under 25 and 58 percent manual workers; 78 percent of the fatalities and 77 percent of the injuries (some 20,000 persons) occurred in Budapest; about half the wounded were under 30.[46] Furthermore, about 200,000 people left the country in those weeks. The fighting in October and November did some 3 billion forints' worth of damage to buildings and production tools and equipment.[47]

According to a report by General Ivan Serov, Soviet troops arrested 4,700 persons after November 4 (6,380 by other counts). Of these, 1,400 were imprisoned, of whom 860 were deported to Ungvár (Uzhgorod) and Sztrij (Stry) in Ukraine.[48] Although the Hungarian government denied in an official communiqué on November 18 that young people and others were being removed to the Soviet Union at all,[49] the truth was that they were only repatriated and their cases handed over the Hungarian authorities after several protests by Kádár and Münnich. The Soviet forces also suffered grave losses: among the 7,349 officers and 51,472 subalterns and men involved there were 2,260 casualties: 85 officers and 584 subalterns and men (669 persons) who lost their lives, 51, including 2 officers, missing, and 1,968 who were injured.[50]

The Hungarian democratic government and society formed by the victorious Revolution were crushed in the war launched by the Soviet Union. The Kádár regime broke all its undertakings as it tried to cow the nation and forestall resistance with one of the bloodiest reigns of terror in contemporary Hungarian history, involving executions, imprisonment, forced exile, deportations and similar methods. The terror lasted until 1961. The Hungarian question was debated by the UN General Assembly, which condemned the Soviet Union in a resolution, despite persistent Soviet efforts to prevent the Soviet aggression against Hungary and the continued Soviet occupation being placed on the agenda. This prompted the

Soviet leadership to order Kádár to reach an agreement with the West and so take the Soviet aggression off the agenda. Kádár obeyed orders, calling a general amnesty, and using international credit to create in Hungary living conditions that led to the country being described as the "happiest barracks" in the Soviet camp and Kádárite society as "goulash communism."

Not until 1989 would the ideas of the Revolution start to apply again in Hungary. A huge aggregate foreign debt had to be repaid under a new, democratic government.[51] One of the first obligations to be met as the negotiations between the Communists and the democratic forces that would replace them got underway was to reinter and ceremoniously rehabilitate Imre Nagy and his fellow martyrs, on June 16, 1989. On the following day, János Kádár died.

Notes

1. The efficiency of Hungarian military intelligence is clear in a report by the military attaché in Paris early in September, which proved a highly accurate prediction: "The plan of attack on Egypt by the Combined Chiefs of Staff would be as follows in the case of armed intervention: preparatory land, air and sea operations by combined French and British forces from the north, from Cyprus, followed by parachute drops to cover the landing of land forces and transport planes. Forces stationed at British bases in Libya and Jordan and combined French troops in Somaliland and Djibouti would go into action simultaneously in the Canal Zone. To support the attack in a southeasterly direction, aircraft carriers would concentrate at the Red Sea entrance, reinforced by New Zealand and Australian forces. The French authorities have mobilized for this end—alongside the Mediterranean Fleet and in addition to the two mentioned divisions—three aircraft carriers, some merchant ships, and colonial, paratrooper and Foreign Legion forces, using similar principles to those of the Indochina expedition force. The main forces are grouping in Toulon, Algiers and Djibouti. There are now 7,000 French and 25,000 British soldiers of mixed branches stationed on Cyprus. If there is an intervention, the French government will call up further reservists." Meanwhile pen-pushers at home took a different view: "According to the assessment of the General Staff of the Hungarian People's Army, it is unlikely that the above military measure will be taken in the present political situation." Hadtörténeti Levéltár, Magyar Néphadsereg [Military History Archive, Hungarian People's Army] (hereafter HL MN) 1956/T–2. ő. e. [2nd guard unit] (hereafter cited as ő. e.) (Sept. 11, 1956).
2. See the Biographies of Key Personalities.
3. On Kádár and Molnár, see the Biographies of Key Personalities.
4. See the Biographies of Key Personalities.
5. Jenő Gergely and Lajos Izsák, *A huszadik század története* [History of the Twentieth Century] (Budapest, 2000), pp. 361–362.
6. See the Biographies of Key Personalities.
7. Gergely and Izsák, *A huszadik század története*, p. 363.

8. See the Biographies of Key Personalities.
9. On Rajk, Pálffy, Szőnyi, Szalai and Sólyom, see the Biographies of Key Personalities.
10. See the Biographies of Key Personalities.
11. See the Biographies of Key Personalities.
12. HL MN 1956/T–2. ő. e. (Feb. 2, 1956).
13. HL MN 1967/T–7–8. ő. e. (April 1, 1956); *ibid.,* T–190/I. ő. e. (April 1, 1956).
14. László Csendes and Tibor Gellért, *Kronológia a Honvédség történetéből 1945–1990* [Chronology of the History of the *Honvédség,* 1945–1990] (Budapest, 1993), p. 159; *Magyarország hadtörténete* [Hungary's Military History], editor-in-chief Ervin Liptai (Budapest, 1985), p. 555.
15. Csendes and Gellért, *Kronológia a Honvédség történetéből 1945–1990,* pp. 161 and 163; *Magyarország hadtörténete,* p. 555.
16. HL MN 1956/T–103/1. ő. e. (June 19, 1956); 01773/ZMKA Pság. (Aug. 25, 1956).
17. Miklós Horváth, *1956 hadikrónikája* [1956 Military Chronicle] (Budapest, 2003), p. 24; Csendes and Gellért, *Kronológia a Honvédség történetéből 1945–1990,* p. 165.
18. HL MN 1956/T–2. ő. e. (Sept. 22, 1956).
19. Csendes and Gellért, *Kronológia a Honvédség történetéből 1945–1990,* pp. 166 and 153–154.
20. On Apró and Mező, see the Biographies of Key Personalities.
21. Lajos Fehér (1917–1981), politician and journalist, joined the Communist Party in 1942 and was prominent in the Budapest partisan groups. He was on the Party's Central Committee in 1956–1981 and the Political Committee in 1957–1975. He headed the political department of the Budapest police as a lieutenant colonel in 1945–1946 before joining the newspaper *Szabad Föld* as an editor (1947–1954). In 1953, he subscribed to Imre Nagy's reform policies. Later a Central Committee secretary (1959–1962), then deputy prime minister, he was responsible for applying the New Economic Mechanism to agriculture in 1966.
22. László Földes (1914–2000), politician, joined the Communist Party in 1937, worked in the Hungarian Front in 1944, and commanded the Újpest partisan group in December 1944–January 1945. He was on the Party Central Committee (1956–1970), head of its Cadre

Department (1956–1957), deputy minister of the interior (1958–1964), deputy minister of agriculture and food (1967–1971) and managing director of Hungexpo (1972–1981).

23. Csendes and Gellért, *Kronológia a Honvédség történetéből 1945–1990*, p. 170.

24. *Ibid.*, p. 155.

25. HM HL Petőfi Sándor Politikai Akadémia (Budapest) 1956-os gyűjtemény [1956 Collection of the Sándor Petőfi Political Academy, Budapest] 1. ő. e.

26. See the Biographies of Key Personalities.

27. Ignác Romsics, *Magyarország története a XX. században* [Hungary's History in the 20th Century] (Budapest, 2001), p. 393.

28. See the Biographies of Key Personalities.

29. Csendes and Gellért, *Kronológia a Honvédség történetéből 1945–1990*, p. 174–178.

30. *Ibid.*, p. 179.

31. *Magyarország hadtörténete*, pp. 559–560.

32. See the Biographies of Key Personalities.

33. *Magyarország hadtörténete*, p. 561.

34. Lee W. Congdon and Béla K. Király, eds., *The Ideas of the Hungarian Revolution. Suppressed and Victorious, 1956–1999* (New York, 2002), pp. 42–63.

35. Csendes and Gellért, *Kronológia a Honvédség történetéből 1945–1990*, p. 178.

36. *Ibid.*, p. 185.

37. *Ibid.*

38. Romsics, *Magyarország története a XX. században*, p. 396.

39. See the Biographies of Key Personalities.

40. See the Biographies of Key Personalities.

41. Béla Király, *Amire nincs ige. Visszaemlékezések, 1912–2004* [For Which There Is No Word. Recollections, 1912–2004] (Budapest, 2004), p. 260.

42. Csendes and Gellért, *Kronológia a Honvédség történetéből 1945–1990*, pp. 193–194.

43. *Ibid.*, p. 196.

44. *Ibid.*, p. 197.

45. Gergely and Izsák, *A huszadik század története*, p. 401.

46. Romsics, *Magyarország története a XX. században*, p. 396.

47. Gergely and Izsák, *A huszadik század története*, pp. 398–399.
48. *Ibid.*, p. 398.
49. Csendes and Gellért, *Kronológia a Honvédség történetéből 1945–1990*, pp. 194–195.
50. Horváth, *1956 hadikrónikája*, p. 443.
51. János Kornai, *Evolution of Hungarian Economy, 1848–1998. II. Paying the Bill for Goulash Communism* (New York, 2000), pp. 232 ff.

Chapter XII
The Miklós Zrínyi Military Academy
in the 1955–1956 Academic Year

The academic year at the Miklós Zrínyi Military Academy began promisingly. Orders of October 20, 1955, from the minister of defense, Colonel General István Bata,[1] confirmed in their posts or transferred to the Academy staff who were to provide a high standard of service to higher military education over a long period.[2]

Major General Endre Matékovits, commander of the Academy,[3] gave orders on November 8, 1955, merging the Special Faculty into the Miklós Zrínyi Academy.[4] (This reversed a move by the minister of defense on September 9, 1955,[5] by among other things abolishing the Ministry of Defense Higher Officers' Course and integrating into the Miklós Zrínyi Academy the Special Course of language training for those being sent to Soviet military academies and the Mobilization—M—Course.)[6] The commander of the Academy subordinated the Special Faculty to himself directly, like the other faculties, appointing Lieutenant Colonel Lajos Móricz as faculty commander and ordering him to report to the studies deputy on the study situation, to liaise with him on future tasks, to prepare for the commencement of the new course, and to instigate the methodological activities required. He also gave orders to the requisite Academy organizations to ensure that the faculty was integrated smoothly. It emerges from the same order that the students were striving for greater recognition, for instance through performance-related pay increases.

Following orders from the minister of defense, Major General Matékovits took action on the end-of-year examinations for the Academy's correspondence students.[7] These were to be held on December 14–30, in Russian, military history, historical materialism, political economy and tactics, and the candidates' branch specialty: artillery firing orders, armored branch studies, military bridge-building or military radio studies.

The commander of the Academy sent the minister of defense a report on November 15, 1955, on the entrance examinations for the first-year 1955–1956 Correspondence Course.[8] Of the 86 called to take them, 76 had done so, 5 had withdrawn, 1 had applied for a year's deferment, and 4 had failed to appear. Another order from the commander of the Academy[9] specified the tasks connected with the completion of the 1954–1955 academic year and commencement of the 1955–1956 academic year. He ordered all faculty commanders and department heads to prepare "specific reports" containing assessments of the last training year and aims and orders for the new training year. The new students were to be received on November 29–30 and teaching would begin on December 1. A separate order[10] governed the conference on November 25 to assess the training year. The plan was to follow that with an account of the minister's order and of the main tasks for 1955–1956, based on the higher commanders' conference.

As preparation for the new training year, the daily routine valid from December 1 onwards was issued on November 28.[11] The working day would last from 08:30 to 17:00. Work lasted until 14:00 on a Saturday and the day before a holiday. The students were woken at 07:00 on weekdays and had six 50-minute lessons from 08:30 to 14:30, with 10-minute recesses between them. Then came an hour for lunch and recreation. Practical activities followed at 15:30–16:15, compulsory private study at 16:30–20:00, handing in of secret materials at 20:00–20:20, supper at 20:20–20:40, individual and club activities at 20:40–23:00, and night rest from 23:00. The difference on a Saturday or the day before a holiday was that reveille was at 06:45, the lessons at 08:00–14:00, lunch and recreation at 14:00–14:40, individual and club activities from 14:40, and lights out at 23:00. On field activities, reveille was at 06:00 and departure at 07:00, with return at 15:30, followed by lunch. Party events could be organized on any day in the first week of each month from 18:00, and political briefings every Tuesday at 16:30–17:30. Obligatory club activities were held every Wednesday at 15:30–17:30.

On December 5, the commander of the Academy submitted to the chief of General Staff a combat strength report of December 1,[12] reporting the following regular/effective strength: generals and officers 320/298, re-enlisted subalterns 522/507, and civilian employees 266/256. So altogether 61 of the regular strength of 1,206 were missing, which can be considered tolerable. The student structure by faculties was as follows:

Year	All-branch	Artillery	Armored	Air Force	Engineers	Signal Troops	Logistics	Total
1st	40	61	-	-	20	19	19	159
2nd	46	35	20	15	13	12	20	161
3rc	41	37	19	16	15	14	15	157
Total	127	133	39	31	48	45	54	477

In addition, there were 16 artillery and 14 armored students on the one-year Higher Commander Course—10 fewer than the regular complement. Of the ÁVH officers, 17 were in the All-branch, 3 in the Signal Troops, and 7 in the Logistics Faculty. So there were 507 army and 27 ÁVH officers starting or resuming their studies at the Academy in 1955–1956.

There was a change in the Command of the Academy in December. The minister of defense appointed Colonel Márton András[13] as general deputy to the commander of the Miklós Zrínyi Military Academy, in an order of December 16, 1955.[14]

On December 21, Major General Endre Matékovits ordered that the end-of-year examinations of the Correspondence Faculty should be combined with those for the opening of the year.[15] This meant that all the first-year and second-year students took end-of-year examinations on January 3–7, with all students being examined in Marxism-Leninism and military history (oral), as well as tactics and the Russian language. In addition, the artillery students took artillery studies, the anti-aircraft students took anti-aircraft firing command, the armored students took tank firing practice and armored technology, the first-year engineers took fortification and the second years mines and detonation barriers, the first-year signal troops took electrical technology and the second years radio technology, and the logistics students took leather and textile technology, motor mechanics and transport organization, as well as fuel technology and food studies.

After the minister of defense had ruled that all those from battalion commander upwards had to learn Russian by the end of the 1958–1959 training year to a level of familiarity with the language of daily life and simpler military expressions and ability to converse on these, the commander of the Academy ordered on January 21, 1956,[16] that the whole corps of officers should take part in language training every Tuesday from 16:15 to 18:15 (with a substitute time of 16:00–18:00 on Friday for those absent for service or health reasons), starting on January 24. However, the order did not apply to those who had learnt Russian as part of their secondary school studies. Those leading the groups had to keep a "Diary of Activity and Progress," and the head of the Department of Foreign Languages and General Studies had to report to the studies deputy by 12:00 each Saturday on how the sessions had gone, who had been absent, and so on. As Major General Matékovits put it in his instruction, "I hope that every officer in the Academy will do all in his power to achieve the progress laid down in the requirements in the Russian language—the learning of which is essential for officers of the army of the vast and ever-growing peace camp headed by the Soviet Union—and obey the order of the Comrade Minister of Defense in this respect in an exemplary fashion."

The Preparatory Course for those going to Soviet academies began on February 6.[17] On March 14, 1956, a year after the institution had taken his name, a statue of Miklós Zrínyi was unveiled there, which exists still.[18] The Academy Command continued to devote attention to ensuring relaxation for the students, who were heavily burdened. Again in 1956, they had a chance to go with their families to army vacation homes (in most cases the one at Balatonkenese) during the April 29–May 8 break.[19]

A new pay system came into force in a minister of defense's order of April 1, 1956.[20] This established 11 pay groups by rank for basic pay and 22 groups according to posts, with Group 1 being the highest group in each case. The commander of the Academy was placed in Group 3 (the same as a chief of section in the General Staff) and his deputies in Group 5 (deputy chiefs of section). The faculty commanders and department heads were in Group 6 (divisional deputy commanders). The commander of the M Course and the department heads for military history and for topography and military geography were in Group 7 (divisional heads of supplies). The commanders of each year, the head of military supplies, the heads of infantry firearms training and physical education, foreign languages and general knowledge, and shooting instruction departments were in Group 8 (General Staff deputy heads of the tactics and training department). The senior teachers and lecturers were placed in Groups 8–13 (branch chiefs of division).

Major General László Hegyi, general deputy to the chief of General Staff, gave orders on June 19, 1956, reorganizing the military academies and officers' schools:[21] "To ensure an improvement in the training and education taking place… in the academies, more rational use of trainer and educator cadres, and more rational maintenance… of the academies," the period of training in the Engineering and Signals Faculties of the Miklós Zrínyi Military Academy was raised to four years, while it remained three in the other faculties, and the students in the current third year would graduate under the old system. Training would start each year on December 1, and officers in three-year faculties would graduate on April 10 each year, while those on four-year courses would graduate on September 30. The Correspondence Faculty of the Miklós Zrínyi Academy would offer only all-branch training from the fall of 1956. Finally, the Ságvári Supplies Officers' School would close and the third-year course would be completed in the Military Supplies Faculty of the Miklós Zrínyi Academy.

So Major General Endre Matékovits issued an order on July 21[22] on the entry examinations for the three-year and four-year basic Academy course. The subjects to be examined orally were social science, basic knowledge of weapons, history, topography, geography and the Russian language, and there were to be written examinations in the Hungarian language and arithmetic, and for anti-aircraft artillery and signals officers only, physics and chemistry. It may well have been worth passing, as admission to the Miklós Zrínyi Academy gave at least temporary protection from the government resolution[23] cutting the strength of the Hungarian People's Army by another 15,000, with effect from December 1.

Debates on freedom and the country's independence taking place in the country in the summer of 1956 had their effects on the Academy too. Officer morale was affected not only by livelihood uncertainties, but by such events as a literary evening held by the Academy's Political Department, where views were heard suggesting "that the policy of the Party was incorrect, and some people even cast doubt on the rectitude of our social system."[24] An Academy visit by Minister of Defense Bata, also in July, was accompanied by a hitherto unthinkable incident at the assembly, described by an eyewitness as follows:

> The minister... himself read out a written question that went like this: "Why don't the leaders who have discredited themselves before the people—above all Comrade Rákosi—resign?" ... Then the minister went red and began to shout:

"Who is this Captain Doma who dares to put this question to me? How dare a captain cite the people?" ... Forcing himself to calm down, the minister said something about it being "a tragedy for the nation if Comrade Rákosi resigned," then, forgetting himself again, raised his voice and called the questioning officer "worm-eaten fruit," voicing a hope that "people like you, Captain, will fall from the tree at the first breeze."[25]

Minister of Defense Bata approved on August 4 a new table of organization to come into force in the fall.[26] This included some significant organizational changes. The lesser changes included renaming the Topography and Military Geography Department simply the Topography Department, the Department of the History of the Art of War the Military History Department, and the Department of Foreign Languages and General Knowledge the General Education Department. The faculties were to change into sections by the end of the year. The new All-branch Section absorbed the Armored Faculty, and the Faculty of Artillery and Anti-aircraft Artillery was split into Artillery and Anti-aircraft sections. The Air Force Faculty and the Correspondence Faculty were abolished. Engineering, Signals and Logistics survived as separate sections. Chemical and Preparatory sections were formed, as was a Military Supply Section, to take over the 100 students transferred to the Academy.

Meanwhile the staff and teaching strength declined from 311 to 298, while the 560 students (including 38 from the ÁVH) fell to 552 (40 ÁVH) plus the 100 trainee military supply officers. One lieutenant generalship and 3 major generalships became a single major generalship, the 35 colonelcies fell to 19, and the 114 majorities to 106. The two lieutenancies were abolished, while the number of lieutenant-colonelcies rose from 136 to 137, of captains from 20 to 30, and of lieutenants from 0 to 5.

The results in the 1955–1956 end-of-year examinations for third-year Miklós Zrínyi Academy students appeared in the "Combined Minutes:"[27]

Dept.	Soviet Party History	Hung. Party History	Party Pol. Activity	Tactics	Artillery Shooting	Anti-aircraft Artillery Shooting	Armored Firing	Signals Technology	Bridge-building	Logistics	Grading Averages
Infantry	4.31	4.25	4.68	4.43							4.42
Infantry	4.18	4.43	4.50	4.43							4.38
Infantry	4.13	4.33	4.53	4.20							4.30
Armored	4.22	4.22	4.44	4.16			4.28				4.24
Artillery	4.23	4.19	4.45	4.38	4.05						4.26
Anti-aircraft	4.31	4.13	4.50	4.50		4.37					4.36
Air Force	4.33	4.55	4.62	4.75							4.56
Engineers	4.21	4.00	4.35	3.92					3.26		3.95
Signals	4.25	4.31	4.00	4.68				4.06			4.26
Logistics	4.45	4.30	4.55	4.55						4.75	4.53
Average	*4.26*	*4.27*	*4.46*	*4.40*	*4.05*	*4.37*	*4.28*	*4.06*	*3.26*	*4.75*	*4.30*

A minister of defense's order of September 22, 1956,[28] appointed Colonel András Márton as commander of the Miklós Zrínyi Military Academy. He had previously been general deputy to the commander.

The results of the 1955–1956 state examinations on September 14–28 were as follows:[29]

Department	Enrollment	Social Studies	Tactics	Russian	Artillery Shooting	Air Defense Artillery Shooting	Signals/ Engineers	Logistics	Grading Averages
C-1 Infantry	16	4.25	4.50	4.25					4.33
C-2 Infantry	16	4.00	4.50	4.31					4.27
C-3 Infantry	17	4.13	4.46	4.13					4.24
C-4 Armored	19	4.05	4.55	3.94					4.18
C-5 Artillery	21	4.09	4.33	4.00	4.24				4.16
C-6 Anti-aircraft	16	4.31	4.56	4.50		4.50			4.47
C-7 Air force	16	4.62	4.68	4.55					4.62
C-8 Engineers	14	3.71	4.00	3.80					3.84
C-9 Signals	16	4.31	4.69	4.00			4.81		4.45
C-10 Logistics	20	4.20	4.45	4.20				4.65	4.37
Total/ Averages	171	4.17	4.47	4.17	4.24	4.50	4.81	4.65	4.29

Colonel András Márton

It can be seen that the averages were very high; nobody received a grade of 2. But events of historical importance left Academy students little time to rejoice.

At 14:00 on October 23, 1956, a delegation from the Petőfi Political Academy arrived in the Political Department of the Miklós Zrínyi Military Academy for talks with the commander of the Academy, Colonel András Márton, his political deputy, Lieutenant Colonel Győző Solti, and faculty representatives. A general assembly was then called, where, just as at the Petőfi Academy, some social science teachers urged identification with the mood of the masses. For instance, it was decided at the meeting to omit from a telegram to be sent to fellow academies in Poland expressions of approval of their conduct in recent Polish events. The Hungarian student demands were accepted in general, and a committee was formed "to work out details further. However, this did not occur, as a quite different situation arose in the meantime."[30]

Minister of the Interior László Piros[31] rescinded an earlier ban and allowed the protest. Lieutenant Colonel Solti therefore saw no reason to forbid Academy personnel to attend. Then, although neither the Ministry of Defense nor the General Staff had called an alert (and this had only happened from the Chief Section for Combat Operations of the General Staff between 19:00 and

19:30),[32] the course of events prompted Colonel Márton to call an alert at the Academy in the evening hours, as civilian groups had arrived several times to demand arms from the Academy authorities, saying "If you are Hungarians, give us arms; the army cannot watch workers being shot at in the city."[33] But the commander of the Academy had resisted the demands and stayed loyal to the existing system.

The Academy was guarded by eight units armed with submachine guns. One participant recalls those early hours as follows:

On arrival at the Academy, I took my submachine gun and reported to Lieutenant Colonel Gyula Zentai, studies deputy to the commander of the Academy, who was energetically arranging matters to do with guarding the Academy territory. Me... he sent to the iron railings by Hungária körút, ordering me to take over command of those already there and prevent strangers from entering Academy land... There were several trucks with banners running along Hungária körút, full of civilians armed with loaded rifles...[34]

The 2nd Battalion of the 15th Armored Regiment at Aszód was sent off to Budapest late in the evening of *October 23* to clear Kossuth Lajos tér of protesters, but news came on their arrival at Cinkota that the size of the crowd made it senseless to go there. The battalion was sent to the Miklós Zrínyi Academy instead, where it arrived at 01:00 on *October 24* to find that "some of the Academy officers had rounded up the soldiers and were encouraging them to ignore the order to fire on the people and to move against the ÁVH at the Radio building..."[35] The commander of the Academy, on the other hand, was acting in the opposite direction, so that the Academy officers received the following orders on October 24 and carried out the following tasks:

- Two officers, with the Academy's armored car and 60 men of the Aszód Battalion, rescued 25 police from the besieged XIVth District Police Headquarters.

- The former Army Supplies School was guarded and the ammunition stored there was transported to the Academy.

- About 80 men guarded the premises of the Communist daily *Szabad Nép*.

- An Academy lieutenant colonel with the armored car of the Academy and with the 12/I–II Mechanized Infantry Battalions from Kecskemét secured the Ministry of Defense officers' married quarters, and another officer group secured the Tolbuhin Barracks, where other officers' families lived.

- The XIIIth District Kontakta Electrical Equipment Factory was secured for a day and a half by 50 officers and 50 men.

A summary written months later, but still reflecting bitterness, described events thus:

> Everyone at the Academy was still behaving relatively calmly. Some of the students were preparing for examinations in the classrooms. But this preparation was not wholehearted, as most of them were discussing political events. The mood in general was that everybody unanimously condemned the 12 years of damaging political and economic tyranny in the Rákosi–Gerő period. They felt immeasurably deceived. Everybody was speculating about where the army leaders were: the minister of defense, the chief of General Staff, the various deputies, and so on. Why weren't they making statements, commanding and leading? Everybody wanted to know what the actual situation was and what stance the People's Army was taking. Sadly, nobody gave an answer. The High Command of the army stayed silent and left the army to itself.[36]

October 25:

- This was the bloodiest day of the Revolution: "During the morning, unarmed demonstrators marched to Parliament. They fraternized with the crews of the Soviet tanks. At about 11 o'clock, special police units opened fire on the crowd from the roof of the Ministry of Agriculture opposite the Parliament and neighboring buildings. Serov, a general of the Soviet secret police, who was watching events, ordered tanks to fire on the peaceful demonstrators. The number of dead exceeded 200.[37]

- After a repeated broadcast call by the minister of defense, 220 officers, 85 subalterns, and 130 men returned to their posts and the Academy could form a battalion.

- Guarded by two machine gun sections, fuel was sent from Zách utca and Csángó utca to the Ministry of Defense, for the use of Soviet tanks.

- After a rumor had spread that a trainload of armed insurgents from Miskolc were coming to the capital, the Engineering Faculty blew up the railway tracks in several places along the embankment beside Könyves Kálmán körút and Pongrác út.

- The commander of the Academy took the ÁVH wounded to the duty medical officers at the institution, where they received first aid. A major, who was in the most critical condition, was transported in the Academy armored car to a temporary Soviet military hospital.

- Covered by two tanks, a group of officers transported arms and ammunition to the Party committees of the Vth, VIth and VIIth Districts and to the Red Star Printing Press, for the use of sub-units on duty there.

- A sub-unit of 120 was provided for the National Center for Monitoring Foreigners (KEOK) and a 50-man sub-unit for the Ministry of the Interior store in Május 1. út.

The Academy obviously supported the Soviet action against the Hungarian revolutionaries in quite an organized way and carried out the orders of the Bolshevist regime, although "the Pol[itical] C[hief Section] and Colonel General Hazai in person sent nobody out to brief the officers. That was requested several times by Colonel Márton, and by Lieutenant Colonels Szentesi and Solti as well." A disproportionate amount of burden and responsibility was placed on a few people at the Group Command, which led to some kind of "upset" on October 25. "Colonel Zólomy, who had been in command of the defense of the Hungarian Radio building on October 23, could no longer get into the building; after that, he became one of the chief duty officers at the M[inistry] o[f] D[efense], but also led the group documenting the situation in Budapest. On returning to his duty station, he blurted out 'No one dares to act here; it seems as if I'm to be the C[hief] of S[taff], I have to take command…, I'm to be the minister and the Chief of Staff as well…',"[38] as he had not received any direction from above.

October 26:

- News came from the Military Transport Department at the chiefs of staff that a train with four truckloads of ammunition and four passenger cars for armed men would be arriving in the capital from Miskolc. So 15 men were sent to Rákos train station, but no such train arrived. They then went across to Kőbánya felső station, where they wanted to stop the fast train from Miskolc, but they thought that it was too late to do that and fired at the train. Although they did not find any ammunition or weapons and they apologized to the passengers, that did not undo the loss of two lives and five wounded.

October 27:

- "Confusion at the Ministry of Defense increased on October 27 and 28. The top leaders were clutching at straws and sending up everything for approval by the Politburo members in the Ministry. Once, Major General Lajos Tóth ordered the commands of the provincial guards to use their weapons if a crowd acted threateningly. This was expressly forbidden by István Kovács, a member of the C[entral] C[ommittee] Politburo, whereupon Major General Tóth… flew into a rage. Conflicting orders were issued left and right, but there was no decisive leadership or measures or information for the troops. On October 27... a detachment of 30–40 men from the Academy occupied the National Theater to protect the *Szabad Nép* offices [opposite]. At the same time, Soviet troops fired on the National Theater, thinking that there were insurgents inside. A view developed in the Chief Section for Operations of the General Staff that the commanders of some units were in cahoots with the insurgents, who somehow learnt of every move being made." Large numbers at the Academy sympathized with the nation and the revolutionaries. This is shown by the way that the Academy personnel endorsed an assembly resolution entitled "The Miklós Zrínyi Military Academy Greets the National Government," calling on Budapest youth to promote peace and creative work and supporting the new government of Imre Nagy.[39] This statement also appeared in the *Szabad Nép*.

- On the other hand, the command continued its support for the regime: the Academy took over the defense of the Council House of the VIIIth District and sent two groups of 100 men each to protect the police stores in Róna utca against the freedom fighters.

- The wife of one officer, a Soviet citizen, was arrested as a spy. Colonel András Márton discovered the woman's whereabouts, released her, and found quarters in the Academy for several officers' wives, to prevent similar occurrences.

- A sub-section was sent out to the MÁVAG engineering factory, from which it kept the Orczy tér/Népszínház utca area under surveillance.

October 28:

- A group of officers sent from the Academy managed to dissuade those preparing to storm the VIIIth District Party committee premises.

- On orders from the commander of the Academy, the Health Service saved the remaining medical instruments, toxic material and drugs from the barracks left unclaimed after the dissolution of the ÁVH; the Signals Section did the same for the signals equipment.

- Academy members gave blood for the Central Hospital.[40]
- This was the day of victory for the Revolution. Imre Nagy's order of a cease-fire was obeyed on both sides; the 16-point program of the Youth began to be implemented. Nonetheless, the Academy leadership tolerated the fact that a group of ten officers sent by the Academy to gather weapons should be involved in several hours' shooting against armed freedom fighters on the corner of Baross utca and Bacsó Béla utca, while the cease-fire was operating well elsewhere.

October 29:

- At 15:00, on the command of Colonel Márton, the Academy personnel paraded at the Academy... Colonel Márton gave a stirring speech and ordered them to board vehicles and go along Hungária körút., Kerepesi út and Rákóczi út, singing as they went, informing them that Hungarian tanks would be waiting for them on the corner of the Nagy körút and Rákóczi út. They were to line up behind these... and proceed to the Kilián Barracks. There they would find waiting for them a Hungarian colonel from the M[inistry] o[f] D[efense], already negotiating with the Soviet troops and the insurgents. As far as we know, the task of the officer group was to... [eliminate the freedom-fighter group.] When the [officer] group arrived at the corner of Rákóczi út and the Nagy körút, Lieutenant Colonel Lajos Teleki... [and his group tried to get into the Kilián Barracks.] On reaching Rákóczi tér, they were stopped by a large crowd of civilians, who told them not to proceed as Soviet troops were strafing the Nagy körút. So Lieutenant Colonel Teleki ordered his fellow officers off the vehicles and lined them up in a narrow side-street opposite. Meanwhile Colonel Márton arrived in a tank and was briefed on the situation by Lieutenant Colonel Teleki. Colonel Márton and his tank headed the group... [and] went to Rökk Szilárd utca. As the group entered the street, they were surrounded by a large number of armed civilians and found machine and submachine guns trained on them from the windows... The Academy group had got into a very difficult situation against the freedom fighters, who were in a position to annihilate them to the last man. According to Lieutenant Colonel Teleki, Colonel Márton told the freedom fighters that he would go and negotiate with the Soviet troops, so that they would retreat from Budapest in line with the cease-fire. If the Soviet sub-unit withdrew, this group would take over the area occupied by them and continue securing it. ... The insurgents were convinced with great difficulty that Colonel Márton would not deceive them... Colonel Márton was away for about three hours... [and returned to] say that agreement had been reached with the Soviet Command. This the insurgents received with great enthusiasm. Colonel Márton asked the crowd if they were prepared... to accept him as their commander. If they were, they should line up behind the officer group. However, this order was not obeyed by the crowd. Instead, they flanked the officer group on each side, as they were still uncertain... Finally, they fell in with much difficulty and the march set off for Kálvin tér. There the group spread out between Liberty Bridge and the Museum and awaited the morning... At about six o'clock in the morning, they boarded vehicles and returned to the Academy...[41]

- Members of the Technical Section were assigned to deliver a Soviet soldier to the József Center, and from there a Soviet first lieutenant, two soldiers and an interpreter to the Soviet-occupied Ministry of Defense. Both tasks were done with great difficulty.
- When the nearby ÁVH barracks were emptied, 26 of the old regime's ÁVH, the agents of terror, received asylum at the Academy. Two Soviet officers were taken back to their command wearing Hungarian uniforms. Also returned safe and sound to their embassy were two Korean officers studying at the Academy.

The Academy under Colonel András Márton had worked with the leaders of the ousted regime and the Soviet forces. But it should not be concluded that most Academy personnel were against the insurgents struggling for democracy and the nation's freedom. In fact, most students and even leading teachers sympathized with Hungarian youth and their demand for radical change. It has been mentioned how the intensity of this feeling was apparent in several events before the Revolution began, for instance in the uproar over the July visit by Minister of Defense Bata, when a student inquired why the leaders—especially Comrade Rákosi—did not resign after discrediting themselves before the people. (In the event, the Communist Party Central Committee did dismiss Rákosi at its meeting on July 18–21.) The same tendency was demonstrated in the likewise mentioned statement in support of the Imre Nagy government adopted at the staff meeting on October 27.

But apart from these and the other previously mentioned events showing the great sympathy of the Academy staff for the Revolution, there are two other facts worth noting. One is that after the Hungarian Revolution and War of Independence had been crushed, most officers refrained from following the lead of their commander, Colonel András Márton, and refused to sign the "Officer's Statement" of loyalty to the puppet Kádár regime. The other event took place on October 29: "The representatives of the People's Army, the police, and the freedom fighters established at the police headquarters in Deák tér the Revolutionary Committee for Public Safety, of which Béla Király became the head."[42] Its task was to turn the freedom-fighter groups fighting independently of each other into the new armed force of the nation, the National Guard, and ensure the consolidation of the reforms of the Imre Nagy government. Béla Király spoke to the author about this.

The organization and leadership of the National Guard required a high command in which the main staff of the commander were expert high-command officers faithful to the principles of the Revolution. I knew many of the Miklós Zrínyi Academy teachers of that period, from the time when I had been the Academy's founding commander and they had been students. I invited half a dozen of these to be leaders of the Operational, Intelligence, Signals, Administrative, Personnel and Logistics Departments. I called on them to bring with them, from among their students, similarly well-trained officers loyal to the Revolution. Within 24 hours, there was an expert staff organizing and commanding the National Guard in Deák tér.

Without the cooperation of the personnel of the Miklós Zrínyi Military Academy it would not have been possible to assemble rapidly an effective leading staff. On the other hand, the enemies of the Revolution spread the smear that "the lightning establishment of the High Command of the National Guard proved that the foreign imperialists had begun to organize the Revolution well before October 1956.

October 30:
- By the early morning hours large crowds had appeared at the Party headquarters in Köztársaság tér... The Academy received orders to send at least 100 people to clear Köztársaság tér. It was reported back from the Academy that only 40–50 could be provided, because otherwise it would leave insufficient strength to defend the Academy if it were attacked... Later the situation became catastrophic; the 40 people sent from the Academy were unable to enter Köztársaság tér because the vast crowd was blocking the side-streets from every side.[43]

Some leaders of the Communist Party and a colonel and a leading political officer from the People's Army had begun counter-revolutionary conspiring in the Köztársaság tér Party building, aimed at overthrowing the Imre Nagy government by using some units of the secret police (ÁVH).[44] The demonstrators set out to occupy the Party building, seeing it as "the counter-revolutionary center of the dissolved ÁVH and the Party apparatus" and "this was done by armed groups after a battle of three hours. The army tanks sent to assist the besieged had defected to the attackers. After the bloody battle, 24 of the captured defenders were lynched... The majority of the 80–100 defenders of the Party building survived the attack and the lynching..."[45]

The counter-revolutionaries condemned this as a crime by the freedom fighters. The lynchings can indeed be condemned insofar as all lynchings can be seen as crimes. Furthermore, here, the lynched men, although they had defended the enemies of Hungarian freedom, had given themselves up, and lynching "prisoners of war" contravenes international law and the ethical norms of the civilized world. Yet whatever readers may think of the incident, objectivity demands a comparison between the two bloodiest days of the Revolution. On October 25, the Bolshevik ÁVH and their Soviet allies massacred 200 peaceful, unarmed demonstrators. On October 30, the armed secret police who were lynched were rightly suspected of defending leaders preparing to crush the victorious Revolution for which the freedom fighters were struggling.

- The commander of the Academy sent ten men to Corvin köz as squad commanders, to organize the armed insurgents into military units. On November 3, they sent home our officers, who did not return on the following day."[46]

- The Academy took over the guard of the ÁVH barracks.

- Colonel Márton sent 40 men to do guard duty at Parliament. They were only replaced by Soviet troops on November 7.

October 31:

- Although in Budapest, the withdrawal of Soviet troops appeared to have begun, the second vast Soviet wave of aggression had actually started on the borders. The first war between socialist states commenced.

November 1:
- In protest, Hungary proclaimed neutrality and quit the Warsaw Pact.
- One witness recalled events thus:
The personnel of the Miklós Zrínyi Academy learnt from the papers and radio that Major General Béla Székely had become their new commander, as Colonel András Márton had been appointed as commander of Budapest's outer defensive ring. Major General Székely was not accepted as commander by the Miklós Zrínyi Academy, which requested Colonel András Márton not to leave the Academy. Several delegations and many former officers of the General Staff waited on Colonel Márton in the first three days of November. Several offered him their services.[47]

- 40 persons from the Academy were ordered to the Ministry of the Interior to disperse the crowd that had gathered there.
November 2:
- Guarding and patrolling of objects in the zone assigned to the Academy continued.
- The Permanent Revolutionary Military Council of the Academy chaired by Lieutenant Colonel János Lipták was formed, with the main task of settling the problems and complaints of the personnel (organizing several-day provincial leave periods, restoring order and discipline at the Academy, ensuring fair distribution of pay, and so on).
November 3:
- The Hungarian army's dismissed officers trickled back into the professional army; the minister of defense, Lieutenant General Károly Janza, issued orders reinstating Lieutenant Colonel Sándor Zsilinszky[48] and Major General Béla Kerekes with effect from November 1.[49]
- At the first of two Revolutionary Military Council meetings, the majority rejected a proposal to remove the political officers and

interpreters from the Academy. The second meeting, mainly on social matters, broke off when József Dudás[50] and his wife appeared, to request help from the Academy in curbing unruly junior leaders in Dudás's group. Although Colonel Márton ordered the arrest of the couple, they were soon released again on Ministry of Defense instructions. Then a delegation from the Corvin Group called on the Council for military assistance, which was refused.

- At an officers' meeting, Colonel Márton spoke up for the employees of the Political Section, the Social Science Department, and the Personnel Department, resulting in a vote of confidence in these. This referred primarily to Lieutenant Colonel Győző Solti, who remained on the staff of the Academy Command. Colonel Márton also briefly reviewed the military aspects of the political situation, announcing that negotiations were taking place with the Soviet Command on the withdrawal of Soviet troops from the country. He said that Hungary's intention was that if the talks were not successful within 24 hours, everyone in the country would rise up in its defense.[51]

- By about midnight, the second wave of Soviet aggression, begun in the early hours of October 31 and amounting to undeclared war on Hungary, was complete, after the Soviet Union had played one of the lowest tricks in modern history. For at noon on November 3, the Soviet and Hungarian delegations meeting in the Parliament building agreed that the Soviet troops would withdraw from Hungary by January 15, 1957. When the Hungarian delegation arrived that evening at the Soviet headquarters in Tököl, with a mandate from the government to sign the morning agreement, all its members were arrested, and at around midnight Soviet troops opened fire on the insurgents, in the first war between two Socialist countries.[52]

November 4:
- At 05:20, Free Kossuth Radio broadcast the following appeal: "This is Imre Nagy speaking, the President of the Council of Ministers of the Hungarian People's Republic. Today at daybreak

Soviet troops attacked our capital with the obvious intention of overthrowing the legal Hungarian democratic government. Our troops are in combat. The government is at its post. I notify the people of our country and the entire world of this fact."[53]

- The Revolutionary Military Council at the Academy held its final meeting, in the presence of its commander, Colonel András Márton. It was decided not to obey the radio call or to oppose the Soviet forces. Colonel Márton ordered the personnel to refrain from all hostile acts. He sent his subordinates into the basements of the buildings, while he as commander and Lieutenant Colonel Győző Solti as his political deputy stood at the main entrance to the Academy, turning their back on the Hungarian nation and awaiting the Soviet troops.

- According to the recollections of an active participant, "On hearing the radio news, I went to Lieutenant Colonel Zentai, who 'entrusted' me with making contact with the Soviet troops expected in Kerepesi út. It was still quite dark when I stopped a Soviet tank at the junction of Hungária körút and Kerepesi út and told those inside that the personnel in the nearby Military Academy did not wish to show any resistance. Then came brief radio communication and the tank on which I was relying turned into Hungária körút and halted in the yard of the Academy."[54]

- Early in the morning, civilians broke into the Ministry of Defense arsenal at the Academy vehicle pool and removed the weapons. On hearing of this, the commander of the Academy again acted as the Soviets wished by ordering the Academy duty officer to prevent further arms from being taken.

- Occupation of the Academy was completed at about 08:00. "The local Soviet commander permitted the officers of the Academy to retain their pistols with 14 rounds of ammunition, but the other weapons and ammunition had to be collected and handed over. They were forbidden to leave the barracks,"[55] according to the summary report. There seems to be a contradiction of Colonel

Lajos Móricz, or he was reporting on a later, altered situation: "In the afternoon, a Soviet infantry unit entered the Academy territory from Zách utca. Their commander seized the arsenal and ordered us to hand over our weapons as well. It was useless for us to prove that we had not been and were not participants in the fighting. That uncertainty caused bitterness in many people. The pessimists were not even ruling out being taken prisoners of war. Several people left the Academy for that reason."[56] Of course, it may be that events have become confused in the author's mind, looking back on them after 50 years, because according to the summary report, "The situation normalized further towards evening. The mood of the Academy officers became calmer, although many were concerned for their families. Academy officers cooperated with the Soviet comrades-in-arms [sic] on guarding the barracks. Movement was curtailed in the Academy, with everyone required to be in their quarters after 19:00." The situation changed on the following day *(November 5)*, when the Academy officers were ordered to hand in their pistols, after a sub-unit passing the Academy had come under fire and its commander considered that the attack had come from the Academy.[57] But in subsequent days, Academy personnel were given increasing responsibility for guarding the barracks without friction occurring between them and the occupying forces.

Colonel Márton dealt in his *November 8* order of the day[58] with tightening order and discipline at the Academy. He ordered the establishment of a guard of 40 with submachine guns. The officers' pistols were returned with 14 rounds of ammunition, but all other ammunition and weaponry had to be handed in to the artillery arms duty chief officer. All clothing materials illicitly removed had to be returned as well. The commanders of each section had to organize complete order in their parts of the buildings between 09:00 and 12:00. They had to ensure that uniform was standardized; even in the yard, a blouse with shoulder straps and a belt had to be worn, along with a cap with the national tricolor, issued on that day instead of the badge showing the national arms. The old system for leaving the barracks was abolished. Personnel could go out of the Academy only in special cases, with the requisite leave.

The Academy orders of *November 10*, 1956, were an expression of thanks. The commander of the Academy's message was as follows:[59]

> Calm and good sense having prevailed in our Academy in the recent difficult days, we have stood firm and united to save the Academy, to guard the wealth of our people and to store inventories of very great worth, and to preserve order and discipline. For this I express my thanks to all commanders, officers, subalterns and civilian employees. At the same time, I call on the entire personnel to act as I do in opposing and condemning those who breach discipline and order... to defend with joint force the unflinching loyalty to our people and high moral standard hitherto characteristic of our Academy, and to end the spread of groundless rumors that upset the morale of our personnel. I expect the whole personnel of our Academy to continue giving support to the strengthening of order and discipline and to carry out my instructions and orders without fail. This is the basis on which our Academy will be able as soon as possible to carry out its work under regular conditions, in line with our goals and calling hitherto.

Based on an appeal by Dr. Ferenc Münnich,[60] deputy prime minister and minister of the armed forces, heard on Hungarian Radio on November 10 and reprinted in the November 11 issue of *Népszabadság*, 2,300 officers registered with the Academy between November 11 and 13. Of these, some later returned to their units, some were demobilized, and some joined the 1st and 2nd Companies of the 2nd Officers' Regiment, a special police terror organization set up to back the Kádár system. The condition for further army service was to sign the Officers' Declaration, but this "deeply disturbed the morale of the Academy's personnel."[61]

On the evening of *November 12*, Colonel Márton held a meeting of department heads, where he "announced that although he did not agree with everything in the declaration, he would nonetheless sign it. (This mainly concerned the "unconditional" agreement and the question of two weeks' pay.)"[62] While he ordered the department heads to make the Declaration known to their subordinates, he said "There is no need to hurry with the decision" (signing).[63] On Márton's proposal, a committee devised recommendations for altering the text of the Declaration that the commander of the Academy handed over to Münnich. On the following day, Münnich sent Major General Gyula Uszta to the Academy and gave a one-day deadline for signing. By December 30, 1956, 108 students and 70 staff officers had been discharged.[64] To that large number of officers refusing to take an oath of loyalty to the Soviet Union's satellite puppet government should be added the many who demonstrated by their absence against the idea of committing themselves to an anti-national regime. Although Colonel Márton signed the Declaration,[65] Uszta in his Army Commander's Order of *November 14* related the following: "The Minister of the Armed Forces of the Hungarian People's Republic has relieved Colonel András Márton of his duties as commander of the Miklós Zrínyi Military Academy at his own request, with immediate effect. Concurrently, on this day, he has entrusted the duties of commander of the Miklós Zrínyi Military Academy to Major General Máté Borbás."

Under that new commander, who doubled as commanding officer of the 2nd Officers' Regiment, members of the Academy who had signed the Declaration performed special police tasks. Instruction was only resumed in the spring of 1957.

Notes

1. See the Biographies of Key Personalities.
2. Hadtörténeti Levéltár, Magyar Néphadsereg [Military History Archive, Hungarian People's Army] (hereafter cited as HL MN) 1961/T–33. őrzési egység [guard unit] (hereafter cited as ő. e.). Miklós Zrínyi Military Academy Commander's Order No. 097 (Nov. 1, 1955).
3. See the Biographies of Key Personalities.
4. HL MN 1961/T–33. ő. e. (Nov. 8, 1955).
5. HL MN 1967/T–163/I. ő. e. (Sept. 22, 1955).
6. According to some—e.g. Lieutenant Colonel Dr. Boldizsár Bakity and Lieutenant Colonel Dr. Antal Mátyási, "Szervezési, hadkiegészítési és igazgatási tanszék" [The Organization, Mobilization and Administration Faculty], *Akadémiai Közlemények. Az akadémia szerveinek története* III/B. *Szárazföldi ágazat tanszékeinek története* 120 (1985) 3: 145.
7. HL MN 1961/T–33. ő. e.. (Nov. 10, 1955).
8. HL MN 1955/T–94/2. ő. e. (Nov. 15, 1955).
9. HL MN 1955/T–33. ő. e. (Nov. 16, 1955).
10. *Ibid.*, Miklós Zrínyi Military Academy Commander's Order No. 0107 (Nov. 23, 1955).
11. HL MN 102/03. 405. ő. e. Miklós Zrínyi Military Academy Commander's Order No. 279 (Nov. 28, 1955).
12. HL MN 1955/T–94. ő. e. (Dec. 1, 1955).
13. See the Biographies of Key Personalities.
14. HL MN 1967/T–38. ő. e. (Dec. 16, 1955); HL MN 1961/T–33. ő. e. Miklós Zrínyi Military Academy Commander's Order No. 0125 (Dec. 30, 1955).
15. HL MN 1961/T–33. ő. e. (Dec. 21, 1955).
16. *Ibid.* (Jan. 21, 1956).
17. *Ibid.* (Feb. 2, 1956).
18. HL MN 102/3. 406. ő. e. Miklós Zrínyi Military Academy Commander's Order No. 57 (March 12, 1956).
19. *Ibid.* Miklós Zrínyi Military Academy Commander's Order No. 62 (March 17, 1956).
20. HL MN 1967/T–190/I. ő. e. (April 1, 1956).

21. HL MN 1956/T–103/1. ő. e. 01773/Miklós Zrínyi Military Academy Command (June 25, 1956).
22. HL MN 1961/T–33. ő. e. (July 21, 1956).
23. HL MN 1967/T–237. ő. e. (July 28, 1956). 154 f. Appendix 4, to General Staff Organization and M. Group Command, No. 00941/1957.
24. Lieutenant Colonel Dr. Jenő Bartos, Lieutenant Colonel (ret.) Lajos Kovács and Lieutenant Colonel (ret.) József Hajdó, "Repülő hadműveleti-harcászati tanszék" [The Air Force Tactics and Strategy Faculty]. *Akadémiai Közlemények. Az akadémia szerveinek története IV. A Légvédelmi és repülőágazat története.* 120 (1986) 4: 29.
25. Sándor Kolozsvári, *A dolgozó népet szolgáltam?* [Did I Serve the Working People?] (Budapest, 2001), p. 72.
26. HL MN 1967/T–169. ő. e. (Aug. 4, 1956).
27. HL MN 1961/T–30. ő. e. 0579/Miklós Zrínyi Military Academy Command (March 20, 1959).
28. HL MN 1967/T–48. ő. e. (Sept. 22, 1956).
29. HL MN 1961/T–13. ő. e. 0426/Miklós Zrínyi Military Academy Command (Feb. 20, 1958).
30. HL Zrínyi Miklós Katonai Akadémia (Budapest) 1956-os gyűjtemény [1956 Collection of the Miklós Zrínyi Military Academy, Budapest] (hereafter cited as HL Zrínyi '56) 1. ő. e., 330 f.
31. See the Biographies of Key Personalities.
32. HL MN 1962/T–11. ő. e. 095/Hdm. Csf-ség (March 12, 1957) 5 f. The role and activity of the General Staff Tactical Group Command in the counter-revolutionary events of Oct. 23–Nov. 4, 1956 (hereafter cited as VK Hadműveleti).
33. *Ibid.* 331 f.
34. *Ibid.*, and Dr. Lajos Móricz, *Békében és háborúban. Egy magyar katonatiszt emlékei a XX. század nyolc évtizedéről* [At Peace and at War: A Hungarian Officer's Memories of Eight Decades of the Twentieth Century] (Budapest, 2000), p. 177.
35. Miklós Horváth, *1956 hadikrónikája* [1956 Military Chronicle] (Budapest, 2003), p. 140.
36. HL Zrínyi '56, 335–336 ff.
37. László Varga, ed., *A forradalom kronológiája és bibliográfiája* [Chronology and Bibliography of the Revolution] (Budapest, 1990), p. 51.

38. VK. Hadműveleti, 10 and 12 ff.
39. Dr. Antal Oroszi, *A Zrínyi Miklós Katonai Akadémia történetének összefoglalása (1947–1996). A magyar katonai vezető- és tisztképzés története (Tanulmánygyűjtemény)* [Summary of the History of the Miklós Zrínyi Military Academy, 1947–1996. The History of the Hungarian Military Leadership and Officer Training (Collection of Studies)] (Budapest, 1996), p. 255.
40. HL Zrínyi '56, 350-351 ff.
41. *Ibid.*, 352–355 ff.
42. Varga, ed., *A forradalom kronológiája és bibliográfiája*, p. 57.
43. HL Zrínyi '56, 16–17 ff.
44. *Ibid.*, 359 ff.
45. Varga, ed., *A forradalom kronológiája és bibliográfiája*, p. 59.
46. HL Zrínyi '56, 360 f.
47. Dr. Lajos Móricz, "A magasabb tisztképzés magyarországi történetének elvi, szervezeti és vezetési kérdései" [Theoretical, Organizational and Leadership Questions of the Hungarian History of Higher Officer Training], *Akadémiai Közlemények* 178 (1991): 23–24.
48. Lieutenant Colonel Sándor Zsilinszky, one of the most competent instructors at the National Defense Academy, was among those accused and convicted in the Béla Király show trial. See Miklós M. Szabó, *A magyar katonai felsőoktatás története 1947–1956* [History of Hungarian Military Higher Education] (Budapest, 2004), p. 98.
49. HL MN 1967/T–48. ő. e. (Nov. 3, 1956).
50. See the Biographies of Key Personalities.
51. HL Zrínyi '56, 366 f.
52. Miklós Horváth, "Soviet Aggression against Hungary: Operations 'Wave' and 'Whirlwind'," in Lee W. Congdon and Béla K. Király, eds., *The Ideas of the Hungarian Revolution. Suppressed and Victorious, 1956–1999* (New York, 2002), pp. 65 ff.
53. Congdon and Király, eds., *The Ideas of the Hungarian Revolution. Suppressed and Victorious, 1956–1999*, p. 528.
54. Móricz, *Békében és háborúban*, p. 182.
55. HL Zrínyi '56, 369 f.
56. Móricz, *Békében és háborúban*, p. 182.
57. HL Zrínyi '56, 369–370 ff.
58. HL MN 102/03. 46. ő. e. (Nov. 8, 1956).

59. *Ibid.* (Nov. 10, 1956).
60. See the Biographies of Key Personalities.
61. HL Zrínyi '56, 373 f.
62. *Ibid.*
63. *Ibid.*
64. *Ibid.* 374 f.
65. *Ibid.*

The Author

MIKLÓS M. SZABÓ (b. 1942)

Professor Lieutenant General (ret.). President of the Miklós Zrínyi Military Academy, Budapest since 1996. Head of Depart-ment of Military History (1981–84) at Miklós Zrínyi Military Academy. Commandant of the János Bolyai Military Technical College (1989–91) and that of the Miklós Zrínyi Military Academy (1991–96).

He got his degree at Teachers' Training College and at the Joint Military College in 1964, at Zrínyi Miklós Military Academy in 1972, in History at Loránd Eötvös University, Budapest in 1976.

Doctor of Philosophy in History in 1977, Doctor of Military Sciences in 1987, corresponding member of the Hungarian Academy of Sciences since 2001.

The fields of his specialization: military operations during World War II; the role of Hungary in World War II; the history of Royal Hungarian Air Force.

He has been organizer as well as participant of great number of conferences in Hungary and abroad.

Member of the Special Military Science Committee of Scientific Qualifying Board of the Hungarian Academy of Sciences (1981–96); Chairman of Military Science Section of the Educational Society of Sciences in Budapest (1982–91); that of Military Sciences Committee for the Hungarian Academy of Sciences (1994–2002); Member of the Social Committee of Hungarian Academy of Sciences (2006).

His awards: "Zrínyi Miklós Memorial Ring" (1987); Middle Cross of the Hungarian Republic, Military Section (1994); "Zrínyi Miklós Prize" (1995); Order for Law Enforcement History (2003); Great Order of the Austrian Republic (2003); Middle Cross of the Hungarian Republic with the Star, Military Section (2004); Degree of Cavalier, Merit for National Security of the French Republic in 2006.

Biographies of Key Personalities

Andropov, Yuri Vladimirovich (1914–1984)

    Soviet politician. Soviet ambassador in Budapest (1954–1957), Andropov later became president of the State Security Authority (KGB) (1967–1982), general secretary of the Communist Party of the Soviet Union and president of the Presidium of the Supreme Soviet (head of state) (1983–1984).

Apró, Antal (1913–1994)

    Housepainter and politician. He joined Hungary's illegal Communist Party in 1931 and was interned or imprisoned seven times. In September 1944 he became a Party Central Committee member and took part in the resistance movement. He was then a Central Committee member continuously from 1945 to 1988. He served as general secretary of the national trade union movement (1948–1952), minister of construction materials (1952–1953), deputy prime minister (1953–1956 and 1956–1971), president of the Patriotic People's Front (1956–1958), minister of construction (October 26–31, 1956), Hungary's permanent COMECON representative (1961–1971), and speaker of Parliament (1971–1984).

Asztalos, János (1918–1956)

    Lieutenant colonel. From October 23, 1956, he was one of those directing the defense of the Budapest Party Committee building in Köztársaság tér. When it was stormed on October 30, he was shot dead and his body defiled by the crowd. He was promoted to full colonel posthumously.

Bata, István (1910–1982)

    Tram driver, politician and army officer. He was promoted to colonel in 1949, then major general and lieutenant general in 1950, and colonel general in 1953. Joining the Communist

Party in 1945, he was a Central Committee member in 1950–1953. He served as chief of General Staff in 1950–1953, then minister of defense until 1956. He lived in the Soviet Union until 1958. In 1959 he was stripped of his rank and took up employment with the Budapest Transport Enterprise.

Beér, János (1905–1966)

Lawyer, university professor and doctor of constitutional and legal studies.

Beleznay, István (1909–1950)

General. After finishing the Ludovika Military Academy he was a captain on the General Staff (1939). In November 1944 he joined the resistance group under Lieutenant General János Kiss. In Debrecen, in February 1945, he became head of the Operations Department at the Ministry of Defense of the provisional government, being promoted to colonel and appointed as head of the Ministry's Training Department in September 1945. Promoted to general in 1947 and appointed as commander of the Székesfehérvár Corps in April 1950, he became a victim of the cult of personality.

Bóka, László (1910–1964)

Writer, literary historian, university professor and corresponding member of the Hungarian Academy of Sciences.

Buzsáki, Géza (b. 1927)

Lieutenant Colonel and head of the planning sub-department at the General Staff Academy (1950–1954).

Demtsa, Pál (1913–1985)

Colonel general. First deputy to the group head of military intelligence.

Dinnyés, Lajos (1901–1961)

Politician. Chair of the National Executive of the Small-holders' Party (FKgP), prime minister (1947–December 1948), and speaker of Parliament from 1958 until his death.

Dudás, József (1912–1957)

Imprisoned several times in the 1920s and 1930s in Romania for illegal Communist activity, he arrived in Budapest in 1940 and joined the Communist and resistance movements. In September 1944 he was a member of the secret delegation under Lieutenant General Gábor Faragho sent to Moscow to seek an armistice in 1944. In 1946 he was arrested in connection with the Hungarian Brotherhood show trial, interned, released in 1948, but then interned again and released only in 1954. During the 1956 Hungarian Revolution, he raised an armed detachment and founded on October 29 the Hungarian National Revolutionary Committee under his own chairmanship. On the same day, his men occupied the offices of *Szabad Nép*, the Communist daily paper, and issued from there a newspaper opposing the government of Imre Nagy. He was dismissed by his own detachment on November 3. Called to Parliament for talks on November 21, he was arrested, sentenced to death by the Military College of the Supreme Court on January 14, 1957, and executed five days later. His conviction was quashed in 1989.

Erdei, Ferenc (1910–1971)

Agronomist, politician, sociographer and member of the Hungarian Academy of Sciences. Winner of the Kossuth Prize. He served as minister of agriculture (1949–1953, 1954–1955) and justice (1953–1954), then deputy prime minister (1955–1956), and as general secretary (1957–1964 and 1970–1971) and vice-president (1964–1970) of the Hungarian Academy of Sciences.

Farkas, Mihály (1904–1965)

Printer and politician. He was a member of the Communist Party Central Committee (1944–1956) and the Politburo (1945–1953 and 1953–1955), and its deputy general secretary (1945–1951); also, as chair of the Military Committee, he ran

Party policy on the military, the police and the political police. As minister of defense (1948–1953), he created the Soviet-style Hungarian People's Army. He became a colonel general in 1949 and a field marshal in 1952. He was among those who initiated and directed the show trials. He was expelled from the Communist Party in 1956 and demoted to private, then was arrested on October 13, and, in the spring of 1957, was sentenced to 14 years' imprisonment. He was released in April 1960 and then worked until his death checking translations for a publisher.

Gerő, Ernő (1898–1980)

Politician. A member of the Central Committee (1944–1956) and the Politburo (1945–1956) of the Communist Party, he became first secretary briefly in 1956. He was a member, along with Mihály Farkas and Mátyás Rákosi, of the "troika" in actual control of the country, a member of the secret Defense Committee and deputy prime minister (1952–1956). He was one of the main persons responsible for the cult of personality, the illegalities and the mistaken economic policies of the earlier 1950s.

Gyulai, Mihály (1912–1966)

Taken prisoner as a battalion commander on July 20, 1944, he worked with the Soviet command on the fourth Ukrainian front until December 31, 1944, then in the provisional Hungarian Ministry of Defense in Debrecen, and then as a battalion, regiment and auxiliary commander. He was a battalion commander at the Kossuth Academy, and after completing the Frunze Academy, commander of the National Defense Academy from December 7, 1951, to November 30, 1953, when he was appointed as deputy commander of the 3rd Artillery Corps. The minister of defense removed him from active service on February 2, 1957, citing "his behavior during the counter-revolution."

Gyurkó, Lajos (1912–1979)

Baker and colonel general, with five years of schooling. After a six-month staff officer course, he was promoted to artillery lieutenant colonel and appointed as commander of the 1st Parachute Battalion in 1949 and the 12th Artillery Division in 1950. After completing the Voroshilov Academy in Moscow in 1951–1954, he became commander of the 3rd Army Corps in Kecskemét. He helped to crush the 1956 Hungarian Revolution. On October 27, he gave orders to fire on protesters entering the prison in Kecskemét and for low-flying fighter planes to strafe demonstrators in Tiszakécske, Csongrád and Szeged. He was dismissed from the army in 1960.

Hegedűs, András (1922–1999)

Politician and sociologist. He was a member of the Communist Party Central Committee (1950–1956), the Politburo (1951–1956), the Executive Committee (1951–1956) and the Central Committee Secretariat (1950–1951), and served as deputy prime minister (1953–1955) and prime minister (1955–1956), signing the document of accession to the Warsaw Pact in May 1955. In 1963 he was put in charge of the Sociological Research Group of the Hungarian Academy of Sciences, but dismissed in 1968 for criticizing the Warsaw Pact invasion of Czechoslovakia. He was expelled from the Communist Party in 1973 for studies critical of the Party state.

Ilku, Pál (1912–1973)

Politician, major general, political divisional chief, deputy minister of defense, and minister of culture during the Kádár era.

Illy, Gusztáv (1909–1950)

Lieutenant general. He passed out as an officer from the Ludovika Academy in 1933 and then studied at the General Staff Academy. He deserted in 1944 and went into hiding, but arrived in Budapest in 1945 with the Debrecen provisional government. He became head of the Personnel Department and then chief of the Training Group when the new defense force was established. He became a lieutenant general and inspector-general of the armed forces in 1949. In April 1950 he was arrested on false charges, condemned to death and executed. He was rehabilitated in 1956.

Janza, Károly (1914–2001)

Lieutenant general. He was deputy minister of defense (1951–1956) and also minister of defense during the 1956 Hungarian Revolution. Stripped of his rank in 1958, he was rehabilitated and restored to the rank of lieutenant general in 1990.

Kádár, János (1912–1989)

Typewriter mechanic and politician, the leader of the Party state from 1956 to 1988. He was on the Central Committee and the Politburo of the Communist Party in 1945–1951 and also on the Executive in 1950–1951. He rejoined the Central Committee and the Politburo in 1956, when he became a Central Committee secretary, then first secretary and then president of the Central Committee Presidium. He was on the Provisional Executive Committee of the new Hungarian Socialist Workers' Party in 1956–1957 and then president of its Central Committee in 1957. Thereafter he was a member of the Central Committee and the Politburo (1957–1988), as well as its first secretary (1957–1985), then general secretary (1985–1988) and finally president (1988–1989). The Rajk and generals' show trials were started while he was minister of the interior (1948–1950), but he too was arrested in 1951 and sentenced to life imprisonment. He was rehabilitated in 1954. He became a minister of state in the Imre Nagy government of 1956 and then prime minister (1956–1958 and 1961–1965) of the so-called Revolutionary Worker-Peasant Government imposed on the country by the Soviet Union. He made no attempt to prevent the conviction and execution of Nagy, despite promising several times to do so.

Kerekes, Béla (1900–?)

Major general, course commander of the National Defense Academy, and founder commander of the Ferenc Rákóczi II Military High School.

Kerekes, Imre (1924–1991)

General. After several high appointments, he took part in the modernization of training in the 1970s as first secretary of the Communist Party Committee at the Miklós Zrinyi Military Academy and a decisive leading figure in the institution.

Király, Béla (b. 1912)

Colonel general. He completed the Ludovika Academy in 1935 and the General Staff Academy in 1942, and served on the eastern front as an officer on the General Staff. After the war, he was the commander of the infantry and then commander of the National Defense Academy. After a show trial, he was condemned to death in 1951, which was commuted to life imprisonment. He was freed in September 1956 and was commander-in-chief of the National Guard in October and November. He then fled to the United States and became a university professor. After the change of regime in 1989–1990, he was a colonel general, a member of Parliament and an external member of the Hungarian Academy of Sciences.

Kovács, István (b. 1911)

Upholsterer and politician. He was first secretary of the Budapest Committee of the Communist Party (1954–October 30, 1956). He became head of the Central Committee's Military Committee on October 23, 1956, and also served on the Military Council of the Ministry of Defense in that month. On October 28 he was taken to Moscow, from where he returned only in August 1958.

Kovács, István (1917–2000)

Head of the Operations Group, then chief of General Staff, and a colonel general. He was arrested on November 4, 1956, at the Soviet military base at Tököl while a member of a

Hungarian delegation. He was sentenced to three years' imprisonment in 1958, but freed under an amnesty in 1962. In 1989 he was rehabilitated and promoted to lieutenant general, then in 1997 to colonel general.

Kozma, István (1896–1951)

Lieutenant general. Commissioned as an infantry lieutenant at the Ludovika Academy in 1915, he fought on the Russian and Italian fronts in the First World War. He commanded a company of the Red Guard under the 1919 Hungarian Soviet Republic. In 1923 he was awarded the title of *vitéz* (knight). He completed the Ludovika Military Academy in 1924–1926 and served in the Ministry of Defense and on the General Staff in 1924–1937. In 1942–1944 he was a major general commanding the Transylvanian Border Guard Forces. Promoted to lieutenant general in July 1944, he commanded the 25th Infantry Division on the Soviet front, but was seriously wounded on July 29. The Arrow-Cross regime appointed him as national commander of the Levente youth movement and National Guard on October 30, 1944, but he never took up the post. He was a US prisoner of war from May to December 1945. He then worked as a laborer in 1945–1947. From April to October 1949, he was commander of the reorganized National Defense Military Academy, but was arrested by the State Security Authority. After a show trial, he was convicted of war crimes and crimes against the people and executed in Budapest. He was later rehabilitated.

Lőrincz, Sándor (1911–1950)

Colonel; group head of logistics/quartermaster.

Maléter, Pál (1917–1958)

Army colonel (1954), major general (October 2, 1956) and colonel general (posthumously in 1990). He started as a medical student, but completed the Ludovika Academy in 1942. In May 1944 he was wounded and taken prisoner by the Soviets, but volunteered for partisans' school, and fought in a partisan detachment in Transylvania in September and October. In 1945 he was commander of the guard battalion protecting the government in Debrecen and then of the Ministry of Defense guard company. He joined the Communist Party in 1945. At the end of the 1940s, he served in the border guards and the guards, then as a corps chief of staff (1950–1952) and in the General Staff. In August 1956 he became commander of the auxiliary military engineer units. He joined the revolutionaries on October 29 and took part in founding the Revolutionary Special Police Committee on October 31, becoming a deputy minister of defense on the same day, and minister of defense on November 3. However, he was arrested on November 4 at Tököl, while negotiating with the Soviet forces. As the fifth accused in the Imre Nagy trial, he was sentenced to death on June 15, 1958, and executed on the following day. He was ceremoniously re-interred on June 16, 1958 and legally rehabilitated on July 6.

Márton, András (b. 1924)

Teacher, then professional army officer. After commanding the Hunyadi Artillery Officers' School (1951–1953) and attending the Voroshilov Academy in Moscow, he became deputy commander of the Miklós Zrínyi Military Academy, and then its commander on October 1, 1956. He was sentenced to ten years' imprisonment for his conduct during the Revolution, but freed under an amnesty on April 3, 1962, rehabilitated, and promoted to lieutenant general.

Matékovits, Endre (1909–1985)

Army officer. He began his studies in October 1927 at the Ludovika Academy, where he became an arms instructor (1939–1943) and then a lecturer at the Firing Range School at Várpalota. In October 1944 he commanded a battalion with the rank of captain, but his battalion did not take the oath of allegiance to Ferenc Szálasi, and he was demoted to commander of a mortar company. He surrendered to Soviet forces on November 4. After time in a prisoner-of-war camp, he became a lecturer at the Kossuth Academy on December 1, 1947, and then served as a staff officer. In 1949–1950 he was commander of the Dózsa Infantry Officers' School, in 1950–1953 head of the Operations Group, and from December 1, 1953, to September 9, 1956, commander of the National Defense Academy, later the Miklós Zrínyi Military Academy. On September 22, 1956, he was appointed as deputy head of the Training Group. He was placed on the reserve list on March 8, 1957.

Merényi-Scholtz, Gusztáv (1895–1950)

Physician; surgeon major general. As a professional military physician, he dealt with medical aspects of military flying and became chief medical officer of the air force in 1936. He organized the first high-altitude research station (1938), where he examined aspects of adaptation to sudden shortages of oxygen. He designed the first Hungarian research institute for flying-related medicine. He published widely and went on several foreign study trips (to Germany, France, the Netherlands and the UK) in the 1930s, dealing with the physiology of flying. He became a full university professor in 1941. From 1940, he belonged to the resistance grouped round Endre Bajcsy-Zsilinszky. Under his command, persecuted people were given refuge. He himself went into hiding before his arrest. In 1945 he became the commander of the Red Cross Hospital in Budapest, and in 1946 surgeon major general and head of the Health Department at the Ministry of Defense. In 1947–1948 he helped to establish the first Hungarian Blood Donor Center and Blood Supply Station. In 1950 he was falsely accused in the Rajk trial, condemned to death and executed, but later rehabilitated.

Mező, Imre (1905–1956)

Originally a tailor's assistant, he joined the Belgian Communist Party in 1929 and the International Brigade in the Spanish Civil War in 1936. He also took part in the French Resistance and prepared the August 17, 1944, Paris uprising as a leader of the Patriotic Militia. He joined the staff of the Greater Budapest Committee of the Hungarian Communist Party in the summer of 1945. He held the post of secretary of the Party's Budapest Committee in 1950–1953 and 1954–1956. He was counted as a supporter of Imre Nagy, arguing on October 23 in favor of permitting the students' demonstration. He was mortally wounded on October 30 while acting as a negotiator during the siege of the Budapest Party Committee building on Köztársaság tér.

Mód, Aladár (1908–1973)

Historian, university professor, writer and doctor of historical studies.

Molnár, Erik (1894–1966)

Historian, economist, philosopher and politician. Elected to the Hungarian Academy of Sciences in 1948 (first as a corresponding member, then a full member in 1949). He was a member of the Central Committee of the Communist Party in 1948–1956 and held various ministerial posts (welfare, information, foreign affairs and justice) between 1944 and 1954, as well as being ambassador in Moscow (1948–1949), president of the Supreme Court (1953–1954), director of the Hungarian Academy of Sciences Institute of Historical Studies (from 1949), president of the Hungarian Historical Society (1956–1964) and twice a winner of the Kossuth Prize (1948 and 1963).

Móricz, Lajos (b. 1923)

General and doctor of military studies. Holder of the Miklós Zrínyi Prize.

Münnich, Ferenc (1886–1967)

Lawyer and politician. He took an officers' examination in 1911. In 1915 he became a prisoner of war in Russia, where he became a member of the Bolshevik Party in 1917 and took part in the civil war in 1918. On his return, he played a part in founding the Hungarian Communist Party. Under the 1919 Hungarian Soviet Republic, he headed the Organization Department of the Military Commissariat, then becoming political commissar to the Budapest Red Guard and the 6th Army Division, and, at the end of June, commander of the Slovak Red Army. He fled to Vienna in August, but worked in Sub-Carpathia in 1920–1921, before arriving in Moscow in 1922. In 1936 he joined in the Spanish Civil War as chief of General Staff of the 15th Spanish Division and later as commander of the 11th International Brigade. In May 1946 he became Budapest chief of police with the rank of police lieutenant general. He then served as ambassador in various places in 1949–1956. On October 24, 1956, he became a member of the Communist Party Central Committee, and served as minister of the interior in Imre Nagy's government on October 26–31, but from November 1956 to January 1958 he was deputy prime minister overseeing the armed forces, then in 1958–1961 prime minister, and from 1961 a minister of state in the second János Kádár government. He was retired in 1965.

Nagy, Imre (1896–1958)

Politician, university professor and member of the Hungarian Academy of Sciences (first as a corresponding member in 1950, then a full member in 1953). He was on the Communist Party Central Committee (1944–1951) and the Politburo (1945–1949, 1951–1955 and 1956), as well as its Presidium (October 28–30) and the Provisional Committee of the Hungarian Socialist Workers' Party (October 30–November 2). He served as deputy prime minister (1952–1953) then prime minister (1953–1955 and 1956). During his first term as prime minister he ended the forced industrialization and violent collectivization of agriculture, closed the internment camps, ended internal exile, and founded the Patriotic People's Front. He was ousted as premier and then from the Central Committee and the Politburo by the Rákosi group in 1955, at which time he became leader of the reform opposition in the Party. He became prime minister again on October 24, 1956, and on November 1 announced Hungary's neutrality and withdrawal from the Warsaw Pact. He took refuge in the Yugoslav Embassy on November 4, when the Soviet forces began to crush the Revolution, but he and others were enticed out on November 22, arrested by a detachment of Soviet officers, and sent to Romania. He was arrested there in April 1957, sent home, tried, and on June 15, 1958, convicted "of initiating and leading efforts to overthrow the people's democratic system of state, and of treason," sentenced to death by hanging, and executed on the following day. He accepted none of the charges and did not request clemency.

Nagy, Tamás (1914–1993)

Economist, doctor of economic sciences, university professor and head of department.

Nógrádi, Sándor (1894–1971)

Electrician and politician. He joined the Communist Party in 1919. He was an official and saw combat in the Red Army under the Hungarian Soviet Republic. He then lived in exile in Czechoslovakia and Romania, and from 1931, in Western Europe, mainly France, was engaged in propaganda work. He organized the Arms for Spain movement from 1936. In 1944 he led a partisan unit in Hungary. He was on the Communist Party Central Committee in 1945–1946, 1957–1962 and 1966–1971. Appointed as political state secretary at the Ministry of Industry in 1945, he became a reserve colonel

(1946), head of the Central Committee agitprop department (1947), political state secretary at the Prime Minister's Office (1947–1948), then political divisional chief of staff in the Hungarian People's Army and deputy minister of defense (1948–1955), as lieutenant general (1948) and colonel general (1955). He took part in the show trials. He headed the Central Committee agitprop department again in the old and new Parties in 1955–1957, then became an ambassador (1957–1960) and finally president of the Central Committee Control Commission (1962–1956).

Otta, István (1909–1972)

Major general. He was commander of the Petőfi Institute for Political Officer Training (1949–1951) and the Stalin Political Officers' Academy (1951–1955), deputy head of the Political Division (1955–1957), then commander of the Ferenc Rákóczi II Military High School (1957–1958) and of the Military History Institute and Museum (1958–1970).

Pálffy, György (1909–1949)

Army officer and member of the Central Committee of the Communist Party (1948–1949). He completed the Ludovika Academy in 1932 and the General Staff Academy in 1937–1939. In 1940 he resigned his staff captaincy and became a manager and then chief executive of the electrical company Egyesült Izzó. He joined the illegal Communist movement in 1942, becoming a member of the Military Propaganda Committee of the Hungarian Front and then heading the Military Committee of the Communist Party. He headed the Military Police Department at the Ministry of Defense in 1945–1947, being responsible for its illegal acts. In 1948 he became inspector-general of the army. Based on false charges connected with the military background to the Rajk trial, he was condemned to death and executed in October 1949. He was partially rehabilitated in 1955.

Pesti, Endre (1920–1991)

Major general. Chief of the Personnel Chief Section of the Defense Ministry, commander of the Combined Officers' School.

Piros, László (1917–2002)

Butcher's assistant. Taken prisoner of war by the Soviets in 1943, he went to a partisans' school before fighting in the Soviet Union and Poland. He became secretary and later deputy general secretary of the Trade Union Council (1945–1949). He was also a member of the Central Committee (1949–1956) and an alternate member of the Politburo (1950–1953 and 1955–1956) in the Communist Party. He was commander of the State Security Border Guard with the rank of major general (1950–1953), deputy minister of the interior (1953–1954) and minister of the interior (1954–1956). From November 1956 he lived in the Soviet Union, returning home in 1958 to work until 1977 as chief engineer and then director of the Szeged Salami Factory.

Pórffy, György (1910–1950)

Army officer. After graduating from the Ludovika Academy, he commanded an artillery company in World War II. Taken prisoner by the Soviets, he volunteered for partisans' school, which he completed in 1944, when he returned with his group to Hungary and took part in several successful actions. In 1945 he commanded the 1st Artillery Division and in September became city commander of Budapest, serving in subsequent years in Pécs and Miskolc, then in 1949 in Budapest again. By then a major general, he was arrested on July 24, 1950, condemned to death on false charges and executed. He was rehabilitated in 1954.

Rajk, László (1909–1949)

Politician. A member of the Central Committee (1945–1949) and the Politburo (1945–1949) of the Communist Party, he served as general secretary of the Hungarian Independence People's Front (1949), minister of the interior (1946–1948) and minister of foreign affairs (1948–1949). He was arrested on May 30, 1949, condemned to death on false charges and executed. He was rehabilitated in 1955.

Rákosi, Mátyás (1892–1971)

Politician. He was a member of the Central Committee and the Politburo of the Communist Party (1945–1956), and its general secretary (1948–1953), then its first secretary (1953–1956). He was also deputy prime minister (1949–1952) and prime minister (1952–1953). He exercised total dictatorship in 1949–1953, and so bore the main responsibility for the many illegalities and the cult of personality in that period. He fled to the Soviet Union in 1956 and died there.

Révay, Kálmán (1911–1950)

Major general. Commander of the Kossuth Academy. After graduating from the Ludovika Academy, he became a field officer in the 1st Hussar Regiment, but was dismissed for political reasons in 1944. He joined the illegal Communist Party in the same year. In November he was arrested as a member of the Liberation Committee of the Hungarian National Uprising and condemned to death, but reprieved and given a ten-year prison sentence. He was taken to Sopronkőhida by the Arrow-Cross and then to the West. On returning home in June 1945, he was engaged as a colonel in organizing the Hungarian army and the Kossuth Academy. In 1946 he became ADC to the minister of defense, then commander of the Kossuth Academy, and, early in 1949, inspector of armored troops in the Ministry of Defense. He was arrested on May 21, 1950, condemned to death on false charges and executed. He was rehabilitated in 1955.

Révész, Géza (1902–1977)

Engineer and colonel general. He joined the Communist Party in 1918 and arrived in the Soviet Union in 1923 under an exchange of prisoners. He became commander of the Kiev partisans' school in 1944, and deputy head of the Cadre Department of the Hungarian Communist Party Central Committee in 1945. He was then ambassador in Warsaw (1947–1948), group head at the Ministry of Defense with the rank of lieutenant general (1948), then chief inspector of the Hungarian People's Army. In 1955–1957 he served as military deputy president of the National Planning Office. Minister of defense in 1957–1960 with the rank of colonel general. Ambassador in Moscow in 1960–1963.

Sólyom, László (1908–1950)

Lieutenant general. He completed the Ludovika Academy in 1931 and the General Staff Academy in 1939. He disagreed with the role that Hungary was prepared to play in World War II, and therefore resigned as a staff captain in August 1941, becoming an employee of the electrical company Egyesült Izzó. He joined the illegal Communist Party in 1942 and took part in 1944 in the work of the Liberation Committee of the Hungarian National Uprising, headed by Lieutenant General János Kiss. In January 1945 he was appointed as Budapest chief of police. He became a military head of division at the Ministry of Defense in May 1946 and chief of General Staff in December 1948. He organized the National Defense Military Academy in 1947 and became its first commander. He was arrested in 1950, sentenced to death on false charges, and executed on August 19. He was partly rehabilitated in 1954 and wholly so in 1990.

Szabó, János (1897–1957)

Fitter. He commanded a company under the 1919 Hungarian Soviet Republic. Resettling from Transylvania to Hungary in 1944, he was a Communist Party member from 1945 until 1949, when he was imprisoned for three months for an attempted illegal border crossing. In 1953 he was falsely accused of spying and held for nine months. On October 25,

1956, he joined the Széna tér group and soon became its leader. They took the ÁVH barracks in Maros utca on October 30 and made it their headquarters, then resisted the Soviet invasion on November 4. He was arrested on November 19, sentenced to death by the Military Council of the Supreme Court on January 14, 1957, without a chance of appeal, and executed on January 19.

Szalai, András (1917–1949)

Turner. Deputy head of the Cadre Department of the Communist Party (1947–1949), he was the fifth accused in the trial of László Rajk, sentenced to death, and executed on October 15, 1949. He was rehabilitated in 1955.

Szalvay, Mihály (1899–1955)

Originally a builder's laborer, he joined the Communist Party in 1917 and fought in the Red Army under the 1919 Hungarian Soviet Republic. Imprisoned after its fall, he escaped to Czechoslovakia and then worked in the labor movement in Austria and Belgium in 1923–1926. He was among the first volunteers for the International Brigade in the Spanish Civil War, becoming a battalion commander. Interned in France, he reached the Soviet Union in 1943 and then fought with the Yugoslav partisans, joining the Hungarian army on his return in January 1945 and reaching the rank of lieutenant general. He was also a member of Parliament from 1954 until his death.

Szeremley, Gyula (1913–1968)

Surgeon captain. He was group head of a military training institute.

Serov, Ivan Aleksandrovich (1905–1990)

Field marshal. As president of the Soviet State Security Committee (KGB) (1954–1958), he directed the state security organizations in Hungary in October and November 1956.

Szőnyi, Tibor (1903–1949)

> Physician. He headed the Cadre Department of the Communist Party Central Committee (1947–1949) and served on the Central and Executive Committees (1948–1949). He was arrested in May 1949 on false charges of spying. His confessions under torture formed the basis for the Rajk trial. He was condemned to death in October 1949 and executed. He was rehabilitated in November 1955.

Stalin (Josif Vissarionovich Dzhugashvili) (1879–1953)

> Politician. Soviet dictator and "generalissimo": general secretary of the Soviet Communist Party and prime minister.

Szűcs, Miklós (1921–1998)

> Colonel. Deputy head of operations of the General Staff in 1956.

Tildy, Zoltán (1989–1961)

> Clergyman of the Reformed Church and politician in the Small-holders' Party (FKgP), which he founded in 1930 with Ferenc Nagy and of which he became executive vice-president. He was expressly anti-German in World War II and defensive of national independence. He helped to establish the Hungarian Historical Memorial Committee in 1942. He advocated an alliance with the Social Democratic Party in 1943 and then joined Endre Bajcsy-Zsilinszky in devising an anti-war manifesto. He served as prime minister (1945–1946) and head of state (1946–1948), and then lived under house arrest until 1956, when he joined the Imre Nagy government as a minister of state from October 26 to November 4. He was arrested in 1957 and sentenced to six months' imprisonment in the Nagy trial. He was freed in 1959 and rehabilitated in 1989.

Tóth, Lajos (b. 1922)

> Major general. After completing the Higher Commanders' Course, he became studies deputy to the commander of the National Defense Academy, which was just being established.

On returning from the Voroshilov Academy in Moscow, he worked in of the Hungarian People's Army in organization and mobilization posts, then as Operations Group chief, acting chief of staff, and general deputy to the chief of General Staff. He was commander of the Miklós Zrínyi Military Academy in 1969–1975.

Uszta, Gyula (1914–1995)

Forest guard. Joining the Communist Party and the partisans in 1943, he filled various Party functions after 1945 and became an army officer in 1948. He was also a member of Parliament, the Presidential Council, the Communist Party Central Committee (from 1957) and the Central Committee Control Commission (from 1966), later holding office in partisans' associations. He served as first deputy minister of defense (1956–1970).

Vas, Zoltán (1903–1983)

Writer and politician. He was secretary of the Economic High Council (1946–1949) and later president of the Planning Office (1953) during Imre Nagy's first term as prime minister. He supported Nagy during the 1956 Hungarian Revolution and was interned in Romania, but was excused from prosecution on compassionate grounds. Thereafter he concentrated on his writing.

Veres, Péter (1897–1970)

Writer and politician. He was president of the National Peasants' Party (1945–1949), minister of construction and public works (1947), then minister of defense. He later chaired the Hungarian Writers' Union (1954–1957).

Zsilinszky, Sándor (1914–1990)

Lieutenant colonel. Department head at the National Defense Academy in 1950–1951.

Selected Bibliography

"Parancsnokság és az akadémiavezető szerveinek története" [The History of the Different Bodies of the Commandership and the Commanders of the Academy]. *Akadémiai Közlemények* 120 (1982) 1.

A Magyar Állam szervei 1944–1950 N–Z [Bodies of the Hungarian State, 1944–1950, N–Z]. Budapest: Közgazdasági és Jogi Könyvkiadó, 1985.

A magyar állam szervei 1950–1970. Központi szervek [Bodies of the Hungarian State, 1950–1970. Central Bodies]. Budapest: Magyar Országos Levéltár, 1993.

Albrecht-Carrié, René. *A Diplomatic History of Europe since the Congress of Vienna.* New York: Harper and Brothers, 1958.

Bakity, Boldizsár and Antal Mátyási, "Szervezési, hadkiegészítési és igazgatási tanszék" [The Organization, Mobilization and Administration Faculty]. *Akadémiai Közlemények. Az akadémia szerveinek története III/B. Szárazföldi ágazat tanszékeinek története.* 120 (1985) 3.

Balogh, Gyula. *A hadiakadémiai képzéstől a szervezett doktori (PhD) képzésig. Tanulmány* [From the Military Academy Training to the Organized Doctoral (PhD) Training. A Study]. Budapest, 2001.

Bartos, Jenő, Lajos Kovács and József Hajdó. "Repülő hadműveleti-harcászati tanszék" [The Air Force Tactics and Strategy Faculty]. *Akadémiai Közlemények. Az akadémia szerveinek története IV. A Légvédelmi és repülőágazat története.* 120 (1986) 4.

Congdon, Lee W. and Béla K. Király, eds. *The Ideas of the Hungarian Revolution. Suppressed and Victorious, 1956–1999.* New York: Atlantic Research and Publications, Inc., 2002.

Congdon, Lee, Béla K. Király and Károly Nagy, eds. *1956: The Hungarian Revolution and War for Independence.* New York: Columbia University Press, 2006.

Csendes, László and Tibor Gellért, *Háborútól a forradalomig 1945–1955. Adalékok a magyar hadsereg történetéből* [From War to Revolution, 1945–1955. Contributions to Hungarian Army History]. Budapest: HM Oktatási és Kulturális Anyagellátó Központ,, 1994.

Csendes, László and Tibor Gellért. *Kronológia a Honvédség történetéből 1945–1990* [Chronology of the History of the *Honvédség*, 1945–1990]. Budapest: HVK Tudományos Munkaszervezési Osztály, 1993.

Djilas, Milovan. *Fall of the New Class.* New York, 1998.

Enzsöl, Gyula. "Rakéta és tüzér tanszék" [The Missile and Artillery Faculty]. *Akadémiai Közlemények. Az Akadémia szerveinek története* III/A. *A Szárazföldi ágazat története.* 120 (1985) 2.

Gergely, Jenő and Lajos Izsák, *A huszadik század története* [History of the Twentieth Century]. Budapest: Pannonica Kiadó, 2000.

Honvédségi Közlöny 39 (April 10, 1947) 12.

Honvédségi Közlöny 7 (1949).

Horváth, Miklós. *1956 hadikrónikája* [1956 Military Chronicle]. Budapest: Akadémiai Kiadó, 2003.

Király, Béla K. *Honvédségből Néphadsereg: személyes visszaem-lékezések, 1944–1956* [From *Honvéd* Army to People's Army: Personal Reminiscences, 1944–1956]. Paris – New Brunswick, NJ: Magyar Füzetek, 1986.

Király, Béla K. *Wars, Revolutions and Regime Changes in Hungary, 1912–2004. Reminiscences of an Eyewitness*, eds. Piroska Balogh, Andrea T. Kulcsár and Tamás Vitek. New York: Columbia University Press, 2005.

Király, Béla K., Barbara Lotze and Nándor F. Dreisziger, eds. *The First War between Socialist States: The Hungarian Revolution of 1956 and Its Impact.* New York: Social Science Monographs – Brooklyn College Press, 1984.

Király, Béla. *Amire nincs ige. Visszaemlékezések, 1912–2004* [For Which There Is No Word. Recollections, 1912–2004]. Budapest: HVG Kiadó Rt., 2004.

Kolozsvári, Sándor. *A dolgozó népet szolgáltam?* [Did I Serve the Working People?]. Budapest: ZMNE, 2001.

Kornai, János. *Evolution of Hungarian Economy, 1848–1998. II. Paying the Bill for Goulash Communism.* New York: Atlantic Studies on Society in Change, 2000.

Lengyel, Ferenc. *A magyar tisztképzés rövid története* [The Short History of Hungarian Officer Training]. Budapest: ZMNE, 2000.

Magyarország hadtörténete (2) [Military History of Hungary II]. Editor-in-chief Ervin Liptai. Budapest: Zrínyi Katonai Kiadó, 1985.

Magyarország történelmi kronológiája [The Historical Chronology of Hungary], vol. IV: *1944–1970.* Budapest: Akadémiai Kiadó, 1982.

Móricz, Lajos. "A Honvéd Akadémia" [The *Honvéd* Academy]. *Akadémiai Közlemények* 194 (1992).

Móricz, Lajos. "A Honvéd Hadiakadémia hallgatója voltam" [I Was a Student of the *Honvéd* Military Academy]. *Honvédségi Szemle* 8 (1983).

Móricz, Lajos. "A magasabb tisztképzés magyarországi történetének elvi, szervezeti és vezetési kérdései" [Theoretical, Organizational and Leadership Questions of the Hungarian History of Higher Officer Training]. *Akadémiai Közlemények* 178 (1991).

Móricz, Lajos. "Adalékok a Zrínyi Miklós Katonai Akadémia történetéhez" [Additional Material on the History of the Miklós Zrínyi Military Academy]. *Honvédelem* 11 (1987).

Móricz, Lajos. *Békében és háborúban. Egy magyar katonatiszt emlékei a XX. század nyolc évtizedéről* [At Peace and at War: A Hungarian Officer's Memories of Eight Decades of the Twentieth Century]. Budapest, 2000.

Mucs, Sándor and Ernő Zágoni, eds. *A Magyar Néphadsereg története 1945–1950* [History of the Hungarian People's Army 1945–1950. Budapest: Zrínyi Katonai Kiadó, 1979.

Nagy, Károly, ed. *Information Bulletin No. 2. Significant Documents of the Hungarian Revolution of 1956*. Budapest: ARP Atlantic Research and Publications Public Foundation, 2006.

Nógrádi. *Új történet kezdődött* [A New Story Begun]. Budapest, 1966.

Oroszi, Antal. "Ötven évvel ezelőtt..." [Fifty Years Ago...]. *Új Honvédségi Szemle* 12 (1997).

Oroszi, Antal. *A Zrínyi Miklós Katonai Akadémia történetének összefoglalása (1947–1996). A magyar katonai vezető- és tisztképzés története (Tanulmánygyűjtemény)* [Summary of the History of the Miklós Zrínyi Military Academy, 1947–1996. The History of the Hungarian Military Leadership and Officer Training (Collection of Studies)]. Budapest: HM Oktatási és Tudományszervező Főosztály, 1996.

Petőfi Szellemében, April 19, 1952.

Petőfi Szellemében, Dec. 1, 1951.

Petőfi Szellemében, Dec. 15, 1951.

Petőfi Szellemében, Jan. 12, 1952.

Petőfi Szellemében, Jan. 16, 1953.

Petőfi Szellemében, Jan. 19, 1952.

Petőfi Szellemében, Jan. 26, 1992.

Petőfi Szellemében, July 14, 1954.

Petőfi Szellemében, July 15, 1950.

Petőfi Szellemében, July 28 1954.
Petőfi Szellemében, July 3, 1952.
Petőfi Szellemében, June 19, 1952.
Petőfi Szellemében, June 30, 1950.
Petőfi Szellemében, June 5, 1952.
Petőfi Szellemében, March 1, 1952.
Petőfi Szellemében, May 1, 1952.
Petőfi Szellemében, May 10, 1952.
Petőfi Szellemében, May 17, 1952.
Petőfi Szellemében, May 20, 1950.
Petőfi Szellemében, May 30, 1950.
Petőfi Szellemében, Nov. 14, 1952.
Petőfi Szellemében, Nov. 17, 1951.
Petőfi Szellemében, Nov. 30, 1952.
Petőfi Szellemében, Nov. 7, 1951.
Petőfi Szellemében, Oct. 20, 1950.
Petőfi Szellemében, Oct. 31, 1952.
Rataj, Endre. "Politikai gazdaságtan és hadigazdaságtan tanszék" [The Political Economy and War Economy Faculty]. *Akadémiai Közlemények. Az akadémia szerveinek története II. kötet Társadalomtudományi ágazat tanszékeinek története.* 120 (1985) 2.
Romsics, Ignác. *Magyarország története a XX. században* [Hungary's History in the 20th Century]. Budapest: Osiris Kiadó, 2001.
Sólyom, László. "Az új Hadiakadémia" [The New Military Academy]. *Honvéd* 12 (1947).
Szabó, Miklós M. *A magyar katonai felsőoktatás története 1947–1956* [History of Hungarian Military Higher Education]. Budapest: Zrínyi Kiadó, 2004.
Sztálin Útján, Aug. 20, 1952.
Sztálin Útján, Jan. 23, 1953.
Sztálin Útján, July 2, 1954.
Sztálin Útján, July 24, 1952.
Sztálin Útján, Nov. 29, 1952.
Sztálin Útján, Nov. 7, 1952.
Sztálin Útján, Sept. 18, 1954.
Sztálin Útján, Sept. 24, 1952.
Sztálin Útján, Sept. 30, 1953.
Varga, József and Lajos Kovács. "Összfegyvernemi harcászati tanszék" [The All-Branch Combat Faculty]. *Akadémiai Közlemények. Az*

Akadémia szerveinek története III/A. *A Szárazföldi ágazat története.* 120 (1985) 2.

Varga, László, ed. *A forradalom kronológiája és bibliográfiája* [Chronology and Bibliography of the Revolution]. Budapest: Századvég – Atlanti Kiadó – 1956-os Intézet, 1990.

Vucinich, Wayne S., ed. *At the Brink of War and Peace: The Tito–Stalin Split in Historic Perspective.* New York: Columbia University Press, 1982.

Name Index

Place Index

Volumes Published in
"Atlantic Studies on Society in Change"

No. 1   *Tolerance and Movements of Religious Dissent in Eastern Europe.* Edited by Béla K. Király. 1977.

No. 2   *The Habsburg Empire in World War I.* Edited by R. A. Kann. 1978.

No. 3   *The Mutual Effects of the Islamic and Judeo-Christian Worlds: The East European Pattern.* Edited by A. Ascher, T. Halasi-Kun, B. K. Király. 1979.

No. 4   *Before Watergate: Problems of Corruption in American Society.* Edited by A. S. Eisenstadt, A. Hoogenboom, H. L. Trefousse. 1979.

No. 5   *East Central European Perceptions of Early America.* Edited by B. K. Király and G. Barány. 1977.

No. 6   *The Hungarian Revolution of 1956 in Retrospect.* Edited by B. K. Király and Paul Jonas. 1978.

No. 7   *Brooklyn U.S.A.: Fourth Largest City in America.* Edited by Rita S. Miller. 1979.

No. 8   *Prime Minister Gyula Andrássy's Influence on Habsburg Foreign Policy.* János Decsy. 1979.

No. 9   *The Great Impeacher: A Political Biography of James M. Ashley.* Robert F. Horowitz. 1979.

No. 10
Vol. I*   *Special Topics and Generalizations on the Eighteenth and Nineteenth Century.* Edited by Béla K. Király and Gunther E. Rothenberg. 1979.

No. 11
Vol. II   *East Central European Society and War in the Pre-Revolutionary 18th Century.* Edited by Gunther E. Rothenberg, Béla K. Király, and Peter F. Sugar. 1982.

No. 12
Vol. III   *From Hunyadi to Rákóczi: War and Society in Late Medieval and Early Modern Hungary.* Edited by János M. Bak and Béla K. Király. 1982.

No. 13
Vol. IV   *East Central European Society and War in the Era of Revolutions: 1775-1856.* Edited by B. K. Király. 1984.

* Vols. no. I through XXXVII refer to the series *War and Society in East Central Europe.*

No. 48	*The Press During the Hungarian Revolution of 1848-1849.*
Vol. XXVII	Domokos Kosáry. 1986.
No. 49	*The Spanish Inquisition and the Inquisitional Mind.* Edited by Angel Alcala. 1987.
No. 50	*Catholics, the State and the European Radical Right, 1919-1945.* Edited by Richard Wolff and Jorg K. Hoensch. 1987.
No. 51	*The Boer War and Military Reforms.* Jay Stone and Erwin A.
Vol.XXVIII	Schmidl. 1987.
No. 52	*Baron Joseph Eötvös, A Literary Biography.* Steven B. Várdy. 1987.
No. 53	*Towards the Renaissance of Puerto Rican Studies: Ethnic and Area Studies in University Education.* Maria Sanchez and Antonio M. Stevens. 1987.
No. 54	*The Brazilian Diamonds in Contracts, Contraband and Capital.* Harry Bernstein. 1987.
No. 55	*Christians, Jews and Other Worlds: Patterns of Conflict and Accommodation.* Edited by Philip F. Gallagher. 1988.
No. 56	*The Fall of the Medieval Kingdom of Hungary: Mohács*
Vol. XXVI	*1526, Buda 1541.* Géza Perjés. 1989.
No. 57	*The Lord Mayor of Lisbon: The Portuguese Tribune of the People and His 24 Guilds.* Harry Bernstein. 1989.
No. 58	*Hungarian Statesmen of Destiny: 1860-1960.* Edited by Paul Bödy. 1989.
No. 59	*For China: The Memoirs of T. G. Li, Former Major General in the Chinese Nationalist Army.* T. G. Li. Written in collaboration with Roman Rome. 1989.
No. 60	*Politics in Hungary: For A Democratic Alternative.* János Kis, with an Introduction by Timothy Garton Ash. 1989.
No. 61	*Hungarian Worker's Councils in 1956.* Edited by Bill Lomax. 1990.
No. 62	*Essays on the Structure and Reform of Centrally Planned Economic Systems.* Paul Jonas. A joint publication with Corvina Kiadó, Budapest. 1990.
No. 63	*Kossuth as a Journalist in England.* Éva H. Haraszti. A joint publication with Akadémiai Kiadó, Budapest. 1990.
No. 64	*From Padua to the Trianon, 1918-1920.* Mária Ormos. A joint publication with Akadémiai Kiadó, Budapest. 1990.

No. 114 *A Millennium of Hungarian Military History.* Edited by
Vol.XXXVII László Veszprémy and Béla K. Király. 2002.

No. 115 *Hungarian Relics. A History of the Battle Banners of the 1848–49 Hungarian Revolution and War of Independence.* Jenő Györkei and Györgyi Cs. Kottra. 2000.

No. 116 *From Totalitarian to Democratic Hungary. Evolution and Transformation, 1990–2000.* Edited by Mária Schmidt and László Gy. Tóth. 2000.

No. 117 *A History of Eastern Europe since the Middle Ages.* Emil Niederhauser. 2003.

No. 118 *The Ideas of the Hungarian Revolution, Suppressed and Victorious, 1956–1999.* Edited by Lee W. Congdon and Béla K. Király. 2002.

No. 119 *The Emancipation of the Serfs in Eastern Europe.* Emil Niederhauser. 2004.

No. 120 *Art of Survival. Hungarian National Defense and Society in Modern Times.* Béla K. Király. Edited by Piroska Balogh and Tamás Vitek. 2003.

No. 121 *Army and Politics in Hungary, 1938–1944.* Lóránd Dombrády.
Vol.XXXVIII Edited by Gyula Rázsó. 2005.

No. 122 *Hungary and the Hungarian Minorities (Trends in the Past and in Our Time).* Edited by László Szarka. 2004.

No. 123 *Roma of Hungary.* Edited by István Kemény. 2005.

No. 124 *National and Ethnic Minorities in Hungary, 1920–2001.* Edited by Ágnes Tóth. 2005.

No. 126 *The Occupation of Bosnia and Herzegovina in 1878.* László
Vol.XXXIX Bencze. 2005.

No. 127 *Wars, Revolutions and Regime Changes in Hungary, 1912–2004. Reminiscences of an Eyewitness.* Béla K. Király. Edited by Piroska Balogh, Andrea T. Kulcsár and Tamás Vitek. 2005.

No. 128 *1956: The Hungarian Revolution and War for Independence.*
Vol. XL Edited by Lee Congdon, Béla K. Király and Károly Nagy. 2006.

No. 129 *The History of the Hungarian Military Higher Education,*
Vol. XLI *1947–1956.* Miklós M. Szabó. 2006.